A great many terrestrial plants live in close association with fungi. The features of this association, which is known as a mycorrhiza, are those of a mutualistic symbiosis. Almost all plants form mycorrhizae whereby the fungus provides soil resources to the plant in exchange for energy provided by the plant. The symbiosis means greater productivity under stress for the plant and a steady energy supply for the fungus.

This book addresses the diverse and complex ways in which mycorrhizae affect plant survival as individuals and populations, and community structure and functioning. An evolutionary/ecological approach is used to describe how and under what conditions mycorrhizae influence basic ecological processes. Applications of these mycorrhizal symbioses range from managing natural and agricultural lands to biotechnological processes that enhance agricultural productivity and sustainability.

The ecology of mycorrhizae will be an invaluable book, applicable to all levels of theoretical and applied research in agronomy, botany, ecology, environmental microbiology, and plant pathology.

The ecology of mycorrhizae

CAMBRIDGE STUDIES IN ECOLOGY

ALSO IN THE SERIES

H. G. Gauch, Jr	*Multivariate Analysis in Community Ecology*
R. H. Peters	*The Ecological Implications of Body Size*
C. S. Reynolds	*The Ecology of Freshwater Phytoplankton*
K. A. Kershaw	*Physiological Ecology of Lichens*
R. P. McIntosh	*The Background of Ecology; Concepts and Theory*
A. J. Beattie	*The Evolutionary Ecology of Ant–Plant Mutualisms*
F. I. Woodward	*Climate and Plant Distribution*
J. J. Burdon	*Diseases and Plant Population Biology*
J. I. Sprent	*The Ecology of the Nitrogen Cycle*
N. G. Hairston, Sr	*Community Ecology and Salamander Guilds*
H. Stolp	*Microbial Ecology: Organisms, Habitats and Activities*
R. N. Owen-Smith	*Megaherbivores: The Influence of Large Body Size on Ecology*
J. A. Wiens	*The Ecology of Bird Communities*
N. G. Hairston, Sr	*Ecological Experiments*
R. Hengeveld	*Dynamic Biogeography*
C. Little	*The Terrestrial Invasion: An Ecophysiological Approach to the Origins of Land Animals*
P. Adam	*Saltmarsh Ecology*
M. F. Allen	*The Ecology of Mycorrhizae*
D. J. Von Willert *et al.*	*Life Strategies of Succulents in Deserts*
J. A. Matthews	*The Ecology of Recently-deglaciated Terrain*
E. A. Johnson	*Fire and Vegetation Dynamics*
D. H. Wise	*Spiders in Ecological Webs*
J. S. Findley	*Bats: A Community Perspective*
G. P. Malanson	*Riparian Landscapes*
S. R. Carpenter & J. F. Kitchell (Eds.)	*The Trophic Cascade in Lakes*
R. J. Whelan	*The Ecology of Fire*
R. C. Mac Nally	*Ecological Versatility and Community Ecology*

The ecology of mycorrhizae

MICHAEL F. ALLEN

Department of Biology, Systems Ecology Research Group, San Diego State University, San Diego, California, USA

Published by the Press Syndicate of the University of Cambridge
The Pitt Building, Trumpington Street, Cambridge CB2 1RP
40 West 20th Street, New York, NY 10011–4211, USA
10 Stamford Road, Oakleigh, Melbourne 3166, Australia

First published 1991
Reprinted 1993, 1996

Printed in Great Britain at the University Press, Cambridge

British Library cataloguing in publication data
Allen, Michael F.
The ecology of mycorrhizae.
1. Mycorrhizae
I. Title
589.20452482

Library of Congress cataloguing in publication data
Allen, Michael F.
The ecology of mycorrhizae / Michael F. Allen.
p. cm. — (Cambridge studies in ecology)
Includes bibliographical references and index.
1. Mycorrhizas. 2. Roots (Botany)—Ecology. I. Title.
II. Series.
QK918.A45 1991 90-36077 CIP
589.2′0452482--dc20

ISBN 0 521 33531 0 hardback
ISBN 0 521 33551 1 paperback

UP

Contents

List of figures

Preface

Mycorrhizal associations represent an enigma to most ecologists. For decades, theoretical ecologists have treated mutualisms, including mycorrhizae, primarily as interesting oddities. Moreover, one of the partners is a fungus, a microorganism, and therefore too small to be seen and bothered with. To see a mycorrhiza, one must look down at the ground instead of admiring the scenery. If anything, the mycorrhizae are either incorporated into the infamous microbial biomass black box (extramatrical hyphae) or as a component of the root biomass (internal mass). A classic example of out-of-sight, out-of-mind.

Despite this oversight, research on the ecology and applied biology of mycorrhizae has continued for over a hundred years and the number of published papers is increasing at an exponential rate. Nevertheless, even many mycorrhizasts (following the terminology of J. L. Harley) have not yet grasped the immense variety of types and roles that mycorrhizae play in both native and human-altered ecosystems. The mycorrhizal fungi are found in every terrestrial ecosystem and may well represent the second largest biomass component of many terrestrial ecosystems.

It is to both groups that this book is addressed with the hopes that it will stimulate a new mutualism between mycorrhizal scientists and ecologists.

Any effort attempting to present an overview of a research field as broad as the ecology of mycorrhizae must necessarily restrict and reduce the emphasis in some areas. For this reason, I have intentionally minimized a general overview of the structure of mycorrhizae, the physiology/biochemistry of mycorrhizae and the applied biology of mycorrhizal ecology. I would refer the reader to a number of outstanding books in these subject areas. I especially refer anyone interested in mycorrhizae to the books by J. L. Harley and the most recent edition by Harley and S. E. Smith (*Mycorrhizal Symbiosis*). These writings inspired my own interest in mycorrhizae as they have for countless numbers of my colleagues.

Several agencies have been extremely helpful in the preparation of this book. The National Science Foundation Ecosystems Studies Program and the United States Department of Agriculture Competitive Grants Program funded most of the research that formed the framework of my ideas. Utah State University Ecology Center provided funds for the literature computer search and the Grey Herbarium Library of Harvard University was especially helpful in finding the obscure and important older references. San Diego State University provided resources to complete the volume.

Over the span of my career, I have been especially stimulated by writings and discussions with a number of ecologists and mycorrhizasts. These especially include E. Allen, N. Stanton, M. Boosalis, D. Coleman, P. B. Tinker, M. Caldwell, J. Richards, J. Tenhunen, R. Virginia and J. Jurinak, as well as G. Safir for our continuing arguments on P as a mechanism for all things mycorrhizal. My guru, James MacMahon, has provided me with special attention by continuing to remind me that all is not mycorrhizal, for encouraging and criticizing my occasional forays into ecological theory, and for providing me with Sal, my Berner Sennen Hund, who patiently helped me through the afternoons of writer's block.

A number of individuals have been especially helpful in the preparation of this volume. Anne Anderson, Carl Friese and Ralph Boerner reviewed the manuscript and made suggestions for improvements. Martha Christensen, my mentor, started my career in mycorrhizal studies and spent considerable hours in reviewing this effort as she has so many others. Jillyn Smith edited the book and made numerous suggestions for improving its clarity. David Read not only reviewed the book, he also nominated me for the task. If the reader wished to place blame for this book, please tell David. I take responsibility for any oversights, omissions, and mistakes within the volume.

Finally, I wish to acknowledge my colleague and wife, Edith, who not only read every word of every draft, but also put up with me during the writing, the re-writing, and all times in between. I would like to dedicate this volume to Bill Maben and my parents, Olin and Donna, and to the hikes in the Rocky Mountains of Colorado and Wyoming that started me on this trail.

1

Introduction

In the early 1880s, Professor A. B. Frank, a distinguished forest pathologist of the Landwirtschaftlichen Hochschule in Berlin, was commissioned by the Minister Für Landwirtschaft, Domänen und Forsten, to undertake a systematic study to promote the production of truffles in Prussia. Although Frank did not succeed in growing truffles, he described the essential structure and functioning of a symbiotic relationship between trees and fungi that he termed a 'mykorhiza,' from the Greek meaning 'fungus-root' (Frank, 1885). The association was mutualistic, he stated, because of the covering of the root by the host fungus and the lack of a detrimental response in the host tree. Moreover, the association was widespread, existing on all individuals observed of several tree species across Europe. Earlier morphological descriptions of these root-inhabiting fungi, Frank's observations, and the insights of Kamienski (1882) on the mutualism between *Monotropa* and its fungal associates, initiated interest in mycorrhizae and their importance in plant survival and production.

Despite the observational and experimental evidence demonstrating that intimate biological associations, such as mycorrhizae, are extremely important and widespread (Boucher, 1985), mutualistic symbioses are often considered as biological oddities, relatively unimportant in ecological and evolutionary processes. Williamson's (1972) paradigm, '[mutualism] is a fascinating biological topic, but its importance in populations in general is small' is widely quoted (e.g. May, 1974). Roughgarden, in his introductory population biology text, states that 'most examples [of mutualism] are tropical' (Roughgarden, 1979). May (1981) also states that there are few examples of mutualisms in temperate climates. May (1974, 1981) suggests that mutualism is mathematically unstable and thus there are few natural examples of importance.

Yet mutualism is an extremely ancient phenomenon in the history of life. Complex life forms began as symbioses of prokaryotic organisms (Margulis & Bermudes, 1985). The integration of mitochondria into another organism to create eukaryotic, single-celled organisms represents the most dramatic change in the biological world since the beginning of life, making possible the present incredible array of life forms. Other intimate organismal interactions resulted in the improved survival of both species and were responsible for initiating life in many habitats. Lichens, mutualistic associations between algae and fungi, are often the initial colonizers on unweathered geological materials. The ability of lichens to weather rock material chemically and physically provides a foundation of unconsolidated material in which plants can later gain a foothold.

Mycorrhiza, a mutualistic symbiosis between plants and fungi, may be one of the most important and least understood biological associations regulating community and ecosystem functioning (Harley, 1971). However, despite the extensive data base amassed since the late 1800s, few discussions of mycorrhizal dynamics have been incorporated into the general ecological literature at the population, community or ecosystem levels and, when it is mentioned, the association is often dismissed as unimportant. For example, most texts on population interactions omit mycorrhizal associations from discussions of coevolution (e.g. Roughgarden, 1979; Merrell, 1981). Barrett (1983) stated '...there is little evidence of the physiological basis of their association [plant–fungus mutualisms]'. Chapin *et al.* (1987) suggested that 'Mycorrhizal hyphae have small-diameter hyphae that increase the surface area of the root system but cost ten percent more to construct than the equivalent mass of roots,' and omitted any discussion of mycorrhizae in their review of plant nutrition. Vogt *et al.* (1986) described the Stark & Jordan (1978) paper on the importance of mycorrhizal mats in tropical forests as '*root mats*' in those [Amazonian] forests took up dissolved nutrients more efficiently than microbes...' (italics mine). Similar statements are commonplace despite the importance of mycorrhizae in such processes as short-circuiting the N-mineralization process in heathlands (e.g. Read, 1983) and the dominant role of mycorrhiza in P nutrition of plants (e.g. Harley, 1971).

deBary (1887) formalized the definitions and differentiated the types of symbioses over a hundred years ago. Symbiosis was simply defined as a state occurring when organisms live in intimate contact. He recognized several types of symbioses including parasitism, commensalism, amensalism, neutralism and mutualism. Symbiotic interactions can be described using +/0/− interactions (Figure 1.1). Although gradients can be seen

SPECIES 2 \ SPECIES 1	+	0	−
+	MUTUALISM	COMMENSALISM	PARASITISM
0	COMMENSALISM	NEUTRALISM	AMENSALISM
−	PARASITISM	AMENSALISM	ANTAGONISM

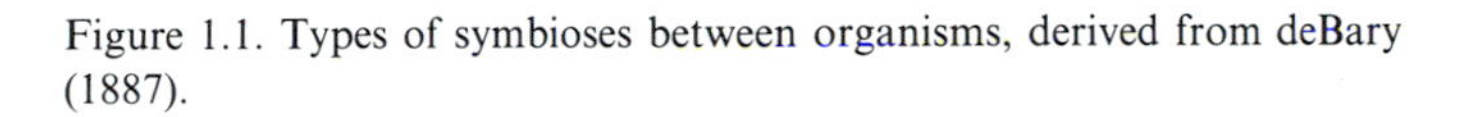

Figure 1.1. Types of symbioses between organisms, derived from deBary (1887).

both evolutionarily (between generations) and in ecological time (within the life of an individual), deBary's definitions provide essential theoretical distinctions.

Although the term mycorrhiza was coined in Frank's 1885 description, the structural and ecological characteristics had been recognized earlier. T. Hartig (1840) clearly illustrated an ectomycorrhiza, and orchid mycorrhizae were described as early as 1851 (see Kelly, 1950). Kamienski described the mutualistic nature of the monotropoid mycorrhiza in 1882 (see the recent translation by S. M. Berch, in Molina, 1985). Frank's (1885) description of a mycorrhiza contains several key elements. These include the mycelial network extending into the substrate as well as the root, and the essential role of the fungus in providing nutrients and water to the plant: '...er functioniert im Bezug auf diese Ernaehrung als die Amme des Baumes' ('it [the mycorrhiza] functions in a nutrition relationship as a wet-nurse of the tree'). The early work was followed by an explosion of descriptions of mycorrhizal associations and arguments regarding the nature of this symbiosis. An increasing and extensive research effort continues today.

Several criteria can be used to distinguish mycorrhizae from other plant–fungus associations. The mutualistic nature of the interaction is a critical character that differentiates a mycorrhiza from other plant–fungus associations. Although the line between parasitism and mutualism is fine (e.g. Harley & Smith, 1983), and negative interactions between plant and fungus can occur both for given species and as environmental conditions change for any one species (e.g. Bethlenfalvay *et al.*, 1982; Buwulda & Goh, 1982; E. Allen & M. Allen, 1984, 1988), the relationship in a general sense is positive for both symbionts (Lewis, 1973). The structural nature of the relationship also distinguishes a mycorrhiza, despite the wide variety of types (discussed in detail in the next chapter); in this relationship, the fungus extends both into the host plant and into the surrounding substrate. Thus, materials move from substrate to host via the fungus. Finally, the relationship is primarily characterized by the flow of inorganic components from fungus to plant and organic components from plant to fungus. Although some experimental data indicate movements in the reverse directions, Lewis (1973) noted that the fungus is heterotrophic and cannot fix significant amounts of carbon, but is structurally and physiologically adapted to move inorganic nutrients (for a detailed discussion on these flows, see Chapter 5). Major exceptions appear to include the protocorm development in orchids (although not necessarily in the mature plant: Alexander *et al.*, 1984; Alexander & Hadley, 1984), and the monotropoid mycorrhizal association in which the fungus may not gain at all (Lewis, 1973). I will use the following definition for my discussion: a mycorrhiza is a mutualistic symbiosis between plant and fungus localized in a root or root-like structure in which energy moves primarily from plant to fungus and inorganic resources move from fungus to plant.

The importance of mycorrhizae is often ascribed simply to the abundance of the association coupled with the perception that if it were not adaptive, the symbiosis would have been selected against. While this argument should not be the sole criterion for importance, the fact of abundance should at least stimulate further interest. Mycorrhizal associations are found in a broad range of habitats. These include ecosystems ranging from aquatic (e.g. Sondergaard & Laegaard, 1977; Bagyaraj *et al.*, 1979) to deserts (e.g. Khudairi, 1969; Williams & Aldon, 1976; Singh & Varma, 1981) and from lowland tropical rain forests (e.g. St John, 1980a; Högberg, 1982; Janos, 1987) to high latitudes (e.g. Malloch & Malloch, 1981, 1982; Christie & Nicolson, 1983; Laursen, 1985), and high altitudes (e.g. Read & Haselwandter, 1981; E. Allen *et al.*, 1987), and in canopy epiphytes (Nadkarni, 1985). Not only are mycorrhizal associations geographically widespread, but within most communities surveyed, mycorrhizae are abundant both within individual root systems and among the array of plant species present. For example, in a semi-arid grassland, all of the dominant plant species had mycorrhizae, and up to 96% of the root length of the dominant species was mycorrhizal (Davidson & Christensen, 1977). In a lowland rain forest, St John (1980a) estimated that about 97% of the 'importance value' of plants represented was mycorrhizal.

Mycorrhizal associations are widespread among plant families and appear to have evolved and spread with the earliest land plants. Stahl (1900) suggested that there are only a few nonmycotrophic families and his designations are generally accepted today (e.g. Gerdemann, 1968; Trappe, 1981). Recent evidence suggests that these 'nonmycotrophic' groups may have mycorrhizal fungal invasions although the physiological relationships are not always defined (Newman & Reddell, 1987). For example, mycorrhizal activity has been reported in the Chenopodiaceae (Williams & Aldon, 1976; M. Allen, 1983), Brassicaceae (Tommerup, 1984; Glenn *et al.*, 1985), *Juncus* and *Carex* (Davidson & Christensen, 1977; Haselwandter & Read, 1980; Christie & Nicolson, 1983; Mejstrik, 1984; E. Allen *et al.*, 1987), *Equisetum* (Lohman, 1927; Koske *et al.*, 1985), *Isoetes* (Farmer, 1988) and hemiparasites (Alexander & Weber, 1985; Lesica & Antibus, 1986), all purported nonmycorrhizal plants. Evidence from the fossil record demonstrates an early appearance of mycorrhizal fungi in roots (Kidstone & Lang, 1921; Wagner & Taylor, 1981; Berch & Warner, 1985; Stubblefield *et al.*, 1987a). These observations, coupled with hypotheses about early terrestrial environments, led to the proposal that invasion of the land by plants depended in part on the evolution of mycorrhizae (Pirozynski & Malloch, 1975; Pirozynski, 1981), which provided essential acids in the acquisition of phosphorus (Fitter, 1985a). Biogeographical evidence also suggests that many of these associations developed early and moved with the plants (e.g. Horak, 1983; Christie & Nicolson, 1983; Halling & Ovrebo, 1987).

To be mutualistic, a relationship must be beneficial to both participants. Growth improvement by host plants in a mycorrhizal association has been demonstrated in studies dating back to Frank (1894). Because many mycorrhizal fungi cannot be grown in pure culture, the fungal symbiont often is presumed to be dependent upon the host plant. The criterion by which mutualism is determined is not a simple matter in all cases. In an agronomic or forestry sense, improved production is generally used as the criterion by which the benefits of mycorrhizae are judged. However, in an ecological context, improved fitness by both symbionts must be the ultimate criterion for judging mutualism, especially in native plants (see the excellent discussion in Lewis, 1973).

Determining the enhanced fitness or survival value of the association over a long time period is difficult or impossible. For example, the giant redwood, *Sequoiadendron giganteum*, is presumed to approach obligate dependence upon mycorrhizal fungi, but no study has maintained a nonmycorrhizal status to maturity and reproduction in the plant. The significance of mycorrhizae to plants is generally based on improved survival of individuals following transplantation into an exotic environment, biomass increases in experimental systems, or altered physiology that can be perceived as an improvement (e.g. increased nutrient uptake or increased drought tolerance). While these approaches are useful, they do not measure fitness. More efforts in this direction are needed.

Numerous studies of the responses of plants transplanted into new habitats have demonstrated the importance of mycorrhizae. Early attempts to establish an exotic forestry industry in Australia, the Caribbean islands, and Rhodesia found that mycorrhizal fungi had to be imported with the plants for those trees to survive (e.g.. Kessell, 1927; Anon., 1931; Hatch, 1936; Briscoe, 1959; Hacskaylo, 1967). Work since the 1930s demonstrated that appropriate mycorrhizal fungi were essential to establishing shelterbelts in grasslands. These include studies from the North American grasslands (e.g. Hatch, 1936; White, 1941; Mikola, 1953; Goss, 1960), and the Ukrainian steppes (see Shemakhanova, 1962; Mishustin, 1967). More recently, experimental transplants and artificial inoculation studies on sterile mine spoil indicate that mycorrhizae improve survival and growth of host plants (e.g. Marx, 1975; Aldon, 1975); Carpenter & Allen, 1988). Survival of conifer seedlings beyond one growing season on the sterile pumice material of the Mount St Helens volcano appeared to require mycorrhizae (M. Allen, 1987a). Improved seed numbers and mass, parameters often related to fitness, have been shown with mycorrhizal establishment several times with different plants (e.g. Carling & Brown, 1980; Yocum, 1983; Carpenter & Allen, 1988). The essential roles of mycorrhizae in orchid culture and *Monotropa* survival have been known since the turn of the century (Kamienski, 1882; Bernard, 1909; Bjorkman, 1960).

The ecological importance of the association to the fungi has been more difficult to prove. In general, the fungus has been presumed to be obligately dependent on the plant because many of the fungi are difficult to culture (see Schenck, 1982). A few field studies suggest that the mycorrhizal fungi have improved activity with a host plant present (e.g. Rommell, 1939; Bjorkman, 1944; E. Allen & M. Allen, 1986).

A major limit to our understanding of mycorrhizal ecology resides in our inability to perceive the scale at which the two symbionts interact with each other and their environment. Often the concepts of population, community, ecosystem and landscape are perceived as linearly increasing size units. This perception has led to difficulties in understanding the ecology of the fungus in particular. An individual fungus may 'perceive' only a few cubic centimeters. Thus, the community of organisms with which it directly interacts may focus along a few centimeters of root length

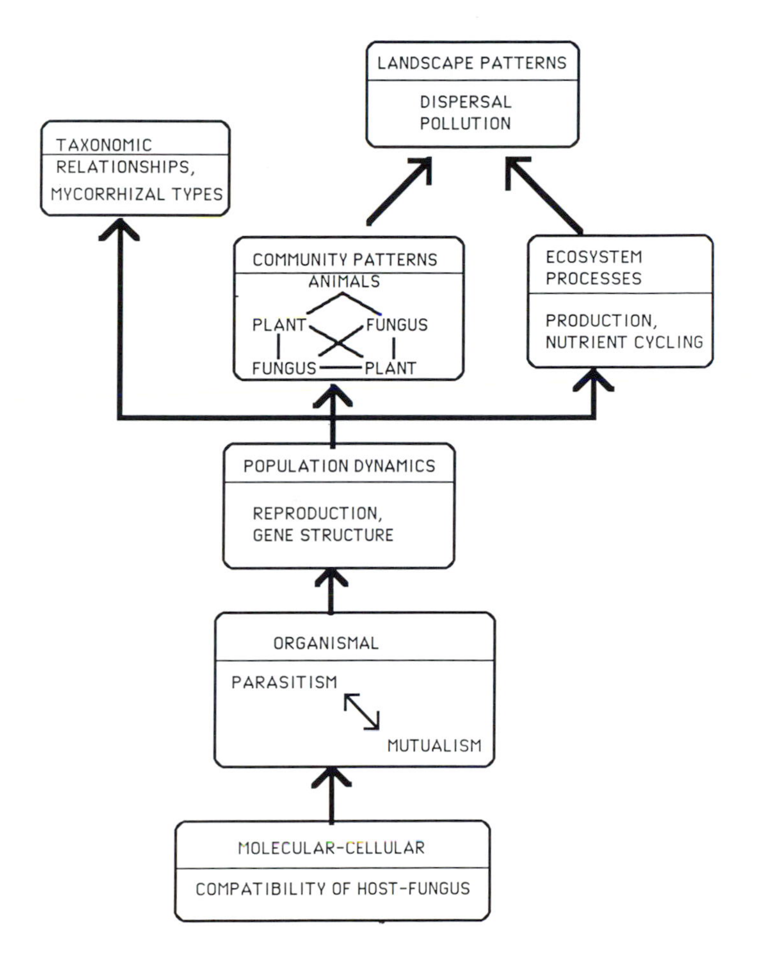

Figure 1.2. Hierarchical approach used to describe the ecology of mycorrhizae. Based on the 'individualist' concept (e.g. MacMahon *et al.*, 1978).

and extend outward only a short distance. In this treatment, I use the hierarchical approach proposed by MacMahon and colleagues (1978) wherein populations, communities and ecosystems are not simply increasing size units but rather are systems that elicit differing types of interactions (see Figure 1.2). With this approach, the relationships of fungus and plant can be distinguished in an ecological, rather than a production agricultural, perspective.

Although the number of mycorrhizasts has increased dramatically in the last few years, the number of research scientists with a basic understanding of mycorrhizae in natural vegetation remains small.

Ecologists, in particular, have ignored mycorrhizal relationships, and this has often led to major difficulties in interpreting ecological data. For example, concentrating only on vertical rooting structure while ignoring the lateral development of mycorrhizal hyphae can lead to gross errors in measuring the spatial distribution of plant nutrient uptake activity (e.g. Parrish & Bazzaz, 1976). Thus a main goal of this book is to initiate an appreciation of the importance of mycorrhizae within the field of ecology in the hope that future ecologists will incorporate an understanding of this symbiosis in their own research.

2

Structure–functioning relationships

The early descriptions of mycorrhizae were based on the morphology of the fungus–host contact zones, the extramatrical hyphae, and the altered root morphologies associated with the symbiosis. The most important feature of the association appears to be the ability of the extramatrical hyphae to take up and transport resources to the plant from the soil outside depletion zones created by the root itself. Overall, knowledge of the structure of mycorrhizal symbioses can dictate an understanding sufficient both for differentiating among the types and for describing the functioning of the association in general. In this chapter, I describe the types of mycorrhizae and their distribution, the structure of the fungal symbiont and the structure of the plant symbiont, and discuss the relationship of the structures to the functioning of mycorrhizae in their environment.

Types of mycorrhizae and their distribution

An early research focus attempted to describe the types of mycorrhizal associations according to symbiont morphology and host–taxon relationships. This focus continues today and is continually being refined with the use of more advanced techniques, especially electron microscopy. As more observations are made, a greater complexity in structure and an increasing diversity of types is encountered. This suggests that a greater emphasis needs to be placed on structure–functioning relationships with an eye toward integrating processes of interest among types as well as toward documenting definitive characteristics among the types of mycorrhizal associations.

Frank (1887, 1891) distinguished two morphological types of mycorrhizae based on whether the fungus penetrated the root cortical cell walls. (N.B. In no reported case does the fungus penetrate the plasmalemma of the host, a common misconception.) He described the two types as endomycorrhizae (penetrating the cell wall) and ectomycorrhizae (no penetration of the cell wall). Other types have been described subsequently, but common practice delineates ecto- (those with the fungus outside the plant cells), ectendo- (those wherein the fungus penetrates the cortical cells but also forms a mantle surrounding the root), and endomycorrhizae (those with hyphae penetrating the cell walls but lacking a mantle). These categories now may be inadequate representations for describing the complex suite of interactions among mycorrhizae.

Wilde *et al.* (1979) and Iyer *et al.* (1980) differentiated endocellular, ectocellular, ectendocellular, and epirhizal mycorrhizal types and suggested that all other descriptions are inadequate. However, these are not really different from the classical groupings except for the addition of the epirhizal form. Harley (1969) discussed in detail the characteristics of a mycorrhiza and noted that there are numerous examples of rhizoplane fungi which might temporarily improve growth. But, as no lasting, coevolutionary relationship could be demonstrated, the rhizoplane fungi (epirhizal) should not be considered as members in mycorrhizal symbioses. The definition of symbiosis (deBary, 1887) clearly implies an intimate, evolutionary relationship, and only when this is shown for rhizoplane fungi, can they properly be referred to as mycorrhizal.

Read (1983) revised the classification system of Lewis (1973) and suggested that mycorrhizal groupings could best be understood in an ecosystem context (Figure 2.1). This system incorporates the types of mycorrhizal associations described to date. In Read's system, the biomes also provide a functioning basis upon which to understand the basic ecology of each relationship. The basic characteristics of the groupings depend upon elevation or latitude gradients, and upon nitrogen and phosphorus availability. As latitude or elevation increases, soils change from mull humus or mineral soil toward a mor soil with a shift from mycorrhizal supply of inorganic nutrients via vesicular-arbuscular (VA) mycorrhizae to the supply of organic nutrients to the host by ericoid mycorrhizae. The ectomycorrhizae are intermediate.

The major limits of Read's distinctions on a biome basis are two-fold. The first takes into consideration the recent work on orchid mycorrhizae by Hadley and colleagues. They demonstrated that in the *Goodyera repens–Ceratobasidium cornigerum* system, carbon was transported from fungus to plant only during the protocorm and plantlet stage (Alexander & Hadley, 1985). In the mature plant, the mycorrhiza acts like most other mycorrhizae, transporting phosphorus to the plant with carbon moving from plant to fungus (Alexander *et al.*, 1984). Orchid and 'saprophytic'

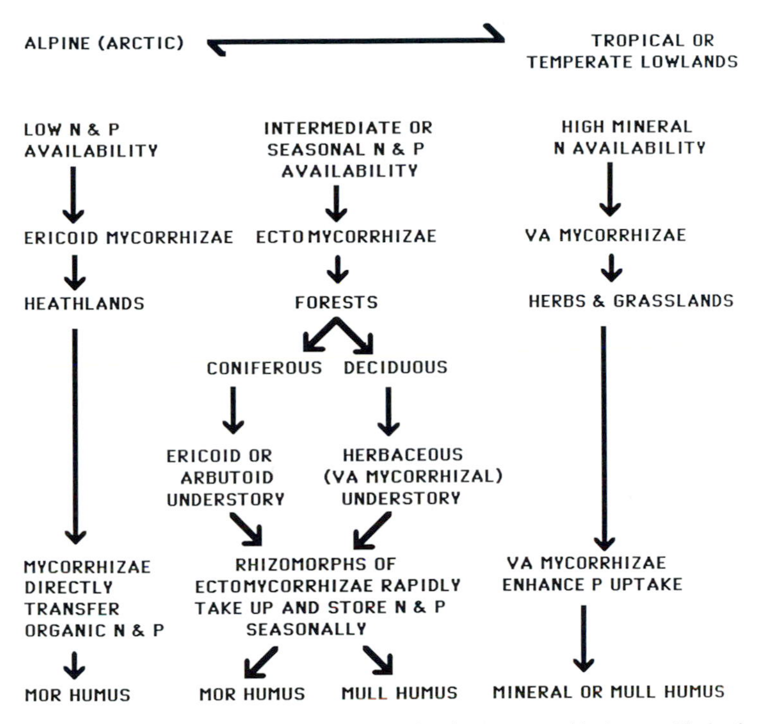

Figure 2.1. Biome types regulating the predominant mycorrhizal types (derived from Read, 1983)

plants (those that receive most, if not all, of their carbon from their mycorrhizal associates and not from humus) are found in virtually every habitat, including soils with little organic matter such as desert soils of the Negev and the high plains of Wyoming (particularly members of the Orobanchaceae), not solely in mor soils. Although more work on a broader range of interactions is needed, these observations suggest that the ecology of mycorrhizae is more complex than Read has depicted.

The second point (usually made for any ecological generalization) is that the types of mycorrhizae and their potential ecological relationships are not yet clear. VA mycorrhizal fungi, for example, have been found both in alpine ecosystems (e.g. E. Allen *et al.*, 1987) and from high latitudes, north of the Arctic Circle in Alaska (M. Allen, unpublished data) and in Antarctica (Freckman *et al.*, 1988). VA mycorrhizae can be abundant in high phosphorous soils (e.g. Davidson & Christensen, 1977; Allen & MacMahon, 1985), in soils beneath mor humus (e.g. Gerdemann & Trappe, 1974; M. Allen *et al.*, 1984b), salt marshes (e.g. E. Allen & Cunningham, 1983; Koh & Lee, 1984), and in canopy epiphytes

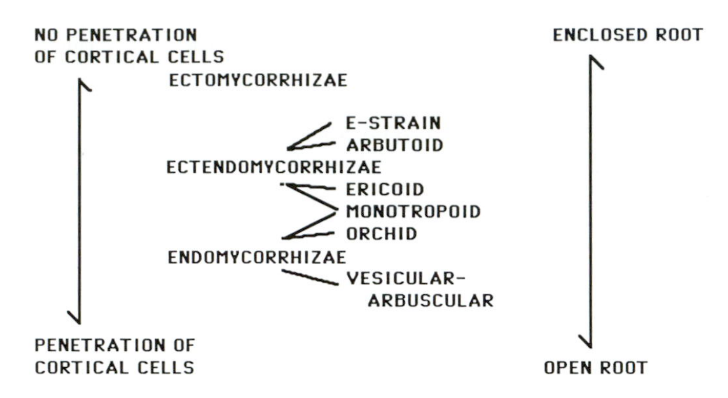

Figure 2.2. Types of mycorrhizae along two structural gradients.

(Nadkarni, 1985). Ericaceous plants and their mycorrhizae are frequently found in inorganic soils such as sand dunes and chaparral (Cowles, 1901; E. Allen & M. Allen, 1990). Thus, the different mycorrhizal types do occur in soils that current dogma regards as atypical. The problem might be addressed by simply stating that the ericaceous mycorrhizae have a greater capacity for using organic nutrient resources and that VA mycorrhizae primarily promote inorganic nutrient uptake. As both resource types are present in many habitats, all mycorrhizal types may be present in any given community type. Establishment, controlled by a range of characteristics at the microhabitat level, is not at the scale of latitudinal gradients, but rather is 'individualistic,' as are plant occurrence patterns (Gleason, 1926).

The types of mycorrhizae might also be viewed as being distributed along two gradients, from an enclosed to open root and from penetration of cortical cells to a lack of penetration (Figure 2.2). This scheme uses characteristics of the fungus–plant interface, consistent with the earlier descriptions. Additional observations and the application of new techniques will improve our understanding of plant–fungus relationships, and eventually will allow a more comprehensive integration of structural, physiological and evolutionary information.

Structure of the fungal symbiont

Although most mycorrhizal treatises focus on structural differences among mycorrhizal types, similarities are probably greater than is widely acknowledged. The roots and internal fungi appear different when observed microscopically; however, all mycorrhizae are composed of an external hyphal matrix and an exchange surface between plant and fungus. Recent evidence suggests that the molecular interactions between plant and fungus at the contact points may be similar between even the two most diverse types of mycorrhizae, ectos and VAs (see Chapter 4). In contrast, much evidence indicates that, physiologically, the fungi are substantially different. The ability to grow the ericoid fungi, for example, on artificial media and their ability to utilize and transport organic nutrients contrasts especially with the VA mycorrhizal types. As Read and colleagues have discussed (e.g. Read, 1983), the capacity to utilize organic substrates may well be a consistent and useful basis for differentiation. I will discuss the structures in a general manner here, and describe physiological functioning in a later chapter.

Probably the most important feature of the mycorrhiza and the most neglected in physiological and ecological research is the external hyphal matrix. Hatch (1937) surmised that the major significance of the mycorrhizal association to the plant was that the hyphae extend beyond the root into the surrounding soil and directly transport nutrients to the plant. Kramer & Wilbur (1949) tested that hypothesis, using ^{32}P, and demonstrated that the external hyphae transport phosphorous from soil to plant. Subsequently, several other workers repeated this type of experiment with other mycorrhizal types and with several other elements (e.g. see Read, 1984; Read *et al.*, 1985). More recently, based upon data from the transport of radiolabeled materials among fungi and plants, researchers have proposed transport of nutrients via the hyphae from one host to fungus to another host, and the transport of organic matter directly from fungus to host (e.g. Woods & Brock, 1964; Herrera *et al.*, 1978; Chiarello *et al.*, 1982; Read, 1984; Read *et al.*, 1985).

The mycelium is not a uniform feature of all fungi but is characteristic of the development within a taxonomic group from which the mycorrhizal fungus is derived. The two major types of external hyphal matrices (Figure 2.3) have important functional differences. The ascomycetous and basidiomycetous mycorrhizal fungi develop virtually all mycorrhizal association types except VA mycorrhizae. These fungi form a 'net' mycelium capable of relatively frequent anastomosis and continuous colonization of a substrate. The zygomycetous mycorrhizal fungi comprise the VA mycorrhizal relationships with the exception of some members of the genus *Endogone* that form ectomycorrhizal associations. The VA mycorrhizal fungi tend to form 'fan'-shaped mycelia consisting of dichotomously branched hyphae radiating out from the trunk or 'runner' hypha (C. Friese unpublished observations). These fungi anastomose infrequently, reducing the frequency of formation of the net mycelia.

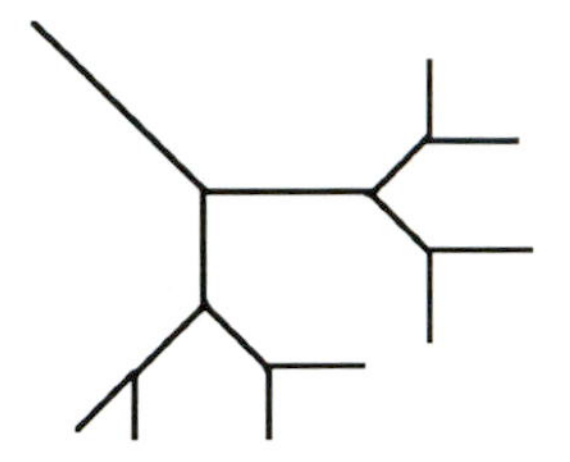

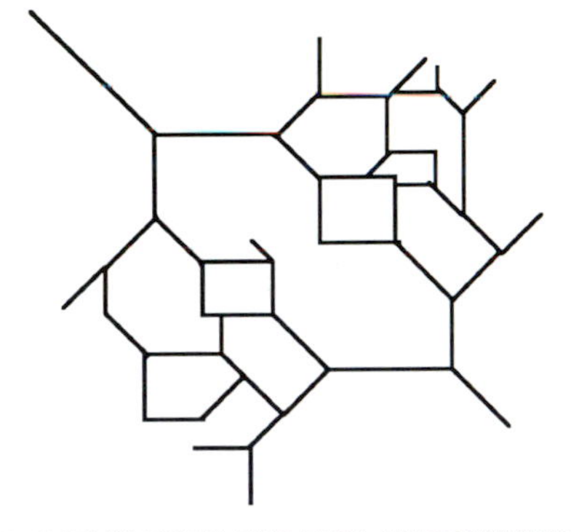

Figure 2.3. A diagrammatic representation of the 'net' versus the 'fan'-shaped mycelium. The net-forming fungi are ascomycetes and basidiomycetes that anastomose between compatible 'individuals'. The fan configuration is formed by dichotomous branching at the hyphal tips, a feature of single colonies of VA mycorrhizal fungi (Zygomycetes) that rarely anastomose.

The net mycelial mycorrhiza-forming fungi exhibit features important to ecosystem dynamics. These have been studied primarily with respect to ectomycorrhizas and the data here are for those systems. Two hyphal types regulate nutrient movement: absorbing hyphae and rhizomorphs. The absorbing hyphae provide the services implied by their name. They are fine, highly branched hyphae that explore substrates, absorbing nutrients released from adjacent soils or organic matter substrates. In some cases, they also secrete external enzymes capable of breaking down organic materials (e.g. Hadley, 1985; Bajwa & Read, 1986; Haselwandter *et al.*, 1987) or otherwise affecting nutrient availability (e.g. Graustein *et al.*, 1977). Rhizomorphs are generally thought of as stress-resisting organs for the fungus. In mycorrhizal fungi, these structures often have an internal conducting structure resembling vascular cylinders in plants, and water and materials flow rapidly through these structures, affecting the

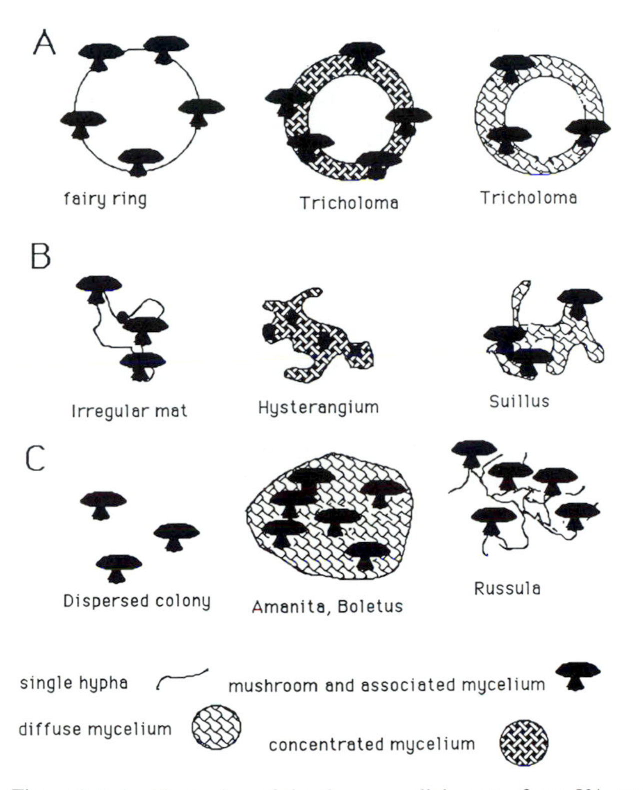

Figure 2.4. An illustration of the three mycelial types of non-VA mycorrhizal fungi, the fairy ring (A), the mat (B), and the diffuse system (C) (from Ogawa, 1985).

water balance of both fungus and host (Duddridge *et al.*, 1980). Hyphal strands are capable of extending far into the soil or even connecting different plants with the same fungus. For example, tree interconnections via mycorrhizal hyphae were hypothesized and experimentally assessed by Woods & Brock (1964) and shown to extend over several meters. The potential ramifications for community ecology and recent evidence testing these assertions will be discussed in Chapter 5.

On a larger spatial scale, the non-VA mycorrhizal associations have three distinctive mycelial structures. All of these fungi form anastomosing networks of hyphae, with the result that compatible fungi, often from different hosts, fuse to form complexes of 'individuals' with differing degrees of cooperation. The extent and significance of this is still being debated in the literature (Chapter 4). Ogawa (1985) differentiated three mycelial forms associated with ectomycorrhizal fungi: a fairy ring, a mat, and a dispersed type (Figure 2.4). A fairy ring mycelium develops just as fairy rings in saprophytic mushrooms develop, as a continually increasing hyphal ring often encompassing several trees. These rings can be quite old. The degree to which exchange of nutrients and genetic material occur still is speculative. The mat-type mycelium occurs as a dominant layer of a single species of ectomycorrhizal mycelium covering a unit of soil. These can often dominate the nutrient cycling characteristics of a soil patch to the exclusion of other fungal species (Cromack *et al.*, 1979). The final type, the diffuse mycelium, is a network that expands in an irregular pattern.

VA mycorrhizal fungi, in the Zygomycetes, have a different developmental pattern. They branch dichotomously with few, adventitious septa; that is, they form few cross-walls between which different cells might gain some measure of independent action. Their growth pattern tends toward a fan-like mycelial structure rather than the 'net' form created as hyphae of different, compatible colonies of ascomycetes or basidiomycetes meet and anastomose. However, endogonaceous fungi have been observed to anastomose more frequently than most zygomycetous fungi with the potential for a greater degree of interaction between individual mycelia than generally perceived. VA mycorrhizal fungi also have two distinct types of extramatrical hyphae, the 'runner' hyphae and the absorbing hyphae (Figure 2.5). The runner hyphae are thick-walled, larger hyphae that track roots into the soil or, in some cases, simply grow through the soil in search of additional roots. The hyphae that penetrate roots are initiated from runner hyphae. The absorbing hyphae also develop from the runner hyphae and form a dichotomously branching hyphal network extending into the soil from the runner hyphae. These hyphae appear to be the component of the fungus that absorbs nutrients from the soil for transport to the host. The distance to which they extend is not well understood. Several reports have suggested that there is *c.* 100 cm of absorbing hyphae per root penetration (e.g. Read, 1984). However, the mechanism limiting hyphal expansion is not known. We have observed runner hyphae extending several centimeters into the soil away from a root. Reports in the literature suggest that the total VA mycorrhizal hyphal length in soil might be as high as 50 m/g but the root density, proportion of root infection, seasonality, and other soil characteristics could influence this value (see for example, M. Allen, 1983; M. Allen & MacMahon, 1985; Caldwell *et al.*, 1985; E. Allen & M. Allen, 1986). VA mycorrhizal hyphae apparently remain relatively close to individual roots, to judge from the branching pattern and P absorption data. Data denoting the major zone of P depletion (Owusu-Bennoah &

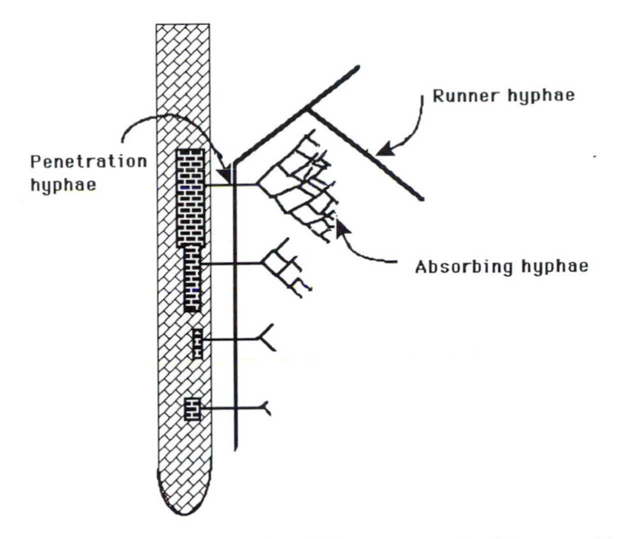

Figure 2.5. An illustration of the structure of a VA mycorrhiza.

Wild, 1979) and aggregation of sand grains (e.g. Koske & Polson, 1984) also support the idea of intensive hyphal development proximate to the root. However, transport up to 7 cm distant has been reported (Rhodes & Gerdemann, 1975).

The second component of the external (to the root) mycorrhiza is the 'reproductive' portion of the fungus. Fungal reproductive structures can be either sexual or asexual. Spores are produced by fungi either for genetic recombination (sexual) or simply as a means of escaping, in space or time, stresses imposed by the environment. In this discussion, I treat the reproductive component as a set of structures for escaping stress. The importance of the reproductive characteristics in relation to the myriad of ways fungi recombine (sexual, parasexual, transfer of extranuclear genes), is discussed in Chapter 4.

Mycorrhizal fungi form a variety of structures for escaping environmental stress. VA mycorrhizal fungi form external chlamydospores or azygospores. The spores are formed either singly or in sporocarps, in loose or tight masses (from one to several hundred) in a hyphal matrix. The other mycorrhizal fungi, primary ascomycetes and basidiomycetes, often form sclerotia, hard stress-resistant masses of hyphae, or can sporulate. Depending on size, vertical distribution, and the environmental characteristics, these structures can be wind, water or animal-dispersed, or they may simply remain quiescent in the soil until conditions adequate for growth are present.

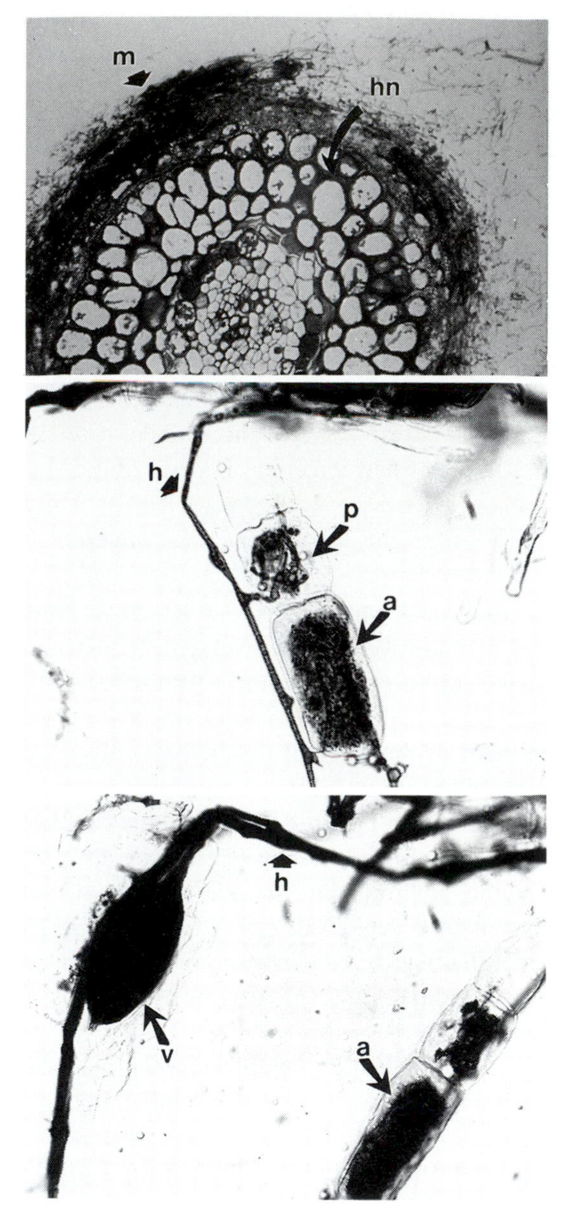

Figure 2.6. Illustrated are the internal structures of the different mycorrhizal types from ectomycorrhizae of *Quercus dumoa* and VA mycorrhizae of *Adenostoma sparsifolia*. Shown are arbuscules (a), vesicles (v), peletons (p), internal hyphae (h), mantle (m), and hartig net (hn).

Two concerns must be expressed regarding the extrapolation from spore densities to estimates of total mycorrhizal activity. First, a number of estimates of mycorrhizal fungal biomass have been computed solely on the mass of these easily discernable structures (see discussion in Read, 1984). The mass of these structures cannot represent the biomass of the entire fungus except possibly during periods of extreme stress when all the external mycelium (the greatest mass component) has died off. Second, spore mass, especially in VA mycorrhizal work, is often assumed to be correlated with inoculum potential. Inoculum potential is a function of host genome × fungal energy × environment × propagule density (see Garrett, 1970). Spores are only one component of propagule density. However, it is an important component, as are seeds to a plant. That is, spores are often the most stress-tolerant and mobile structures of the fungus, important in initiating new infections in new habitats or in highly variable environments.

As the fungus approaches the root surface, a variety of structures become important, depending on the type of mycorrhiza. The interface features basically can be divided into three structural units, the mantle, the hartig net, and the arbuscule. Other structures are present, including the internal hyphae (connecting the external absorbing hyphae with the transfer structures), the vesicles, fungal storage units, and peletons, which may function similarly to arbuscules but for which there are few data. These structures are illustrated in Figure 2.6.

Internal structures are probably the best-described components of the mycorrhiza. They are the organs wherein nutrients and carbon are exchanged between host and endophyte. They also form the basis of the original distinctions between ecto- and endomycorrhizae. Ectomycorrhizae are composed of the mantle and hartig net, fungal exchange structures external to the cortical cells of the host plant. In endomycorrhizae, arbuscules or peletons are formed within the cortical cells and are the exchange organs. Although the fungus penetrates the cell wall, it does not penetrate the membrane. The plasmalemma surface area of a plant is dramatically increased when the plant is mycorrhizal, substantially increasing the surface area for absorption of nutrients compared with the absorptive area of a nonmycorrhizal plant.

At exchange interfaces, nutrients are converted from forms designed for internal transport in one organism to forms more readily utilized and protected in the other. For example, glucose and sucrose, sugars readily utilized by the plant, are taken up by the fungus and converted to trehalose and mannitol, sugar forms readily utilized and transported by the fungus but which the plant generally cannot use (e.g. Lewis & Harley, 1965). Phosphorus appears to be transported through the fungus in polyphosphate granules, possibly in microbody-like structures (e.g. Cox & Sanders, 1974; Cox & Tinker, 1976). At the interface, alkaline phosphatases are concentrated (Bartlett & Lewis, 1973; Gianinazzi-Pearson & Gianinazzi, 1983) and these granules are broken down, releasing locally high concentrations of inorganic phosphate. In VA mycorrhizae, there is a decreasing concentration gradient in phosphate concentration from arbuscule through the xylem sap (Schoknecht & Hattingh, 1976). As the rate of photosynthesis often increases in mycorrhizal plants, increasing the conversion rates of inorganic phosphate to organic phosphate, this pathway actually decreases the free energy necessary to transport phosphate in the host and has been hypothesized to be important in the increased P uptake associated with mycorrhizal infection (M. Allen *et al.*, 1981a). For details on the structure of host–fungus interfaces, I recommend to the reader the outstanding reviews by Harley & Smith (1983), several chapters in Schenck (1982), and Bonfante-Fasolo *et al.* (1984).

Structure of the plant symbiont

Because the mycorrhiza is not just the fungus but the fungus plus the root, understanding the entire structure is important in attempting to understand the ecology of mycorrhizae. The structure of an ecto-mycorrhiza is clearly different from the structure of a nonmycorrhizal root. Frank (1885) described the enclosure of the short roots by the fungus in his original discussion. Short roots are branches from a main root that accommodate formation of mycorrhizae. Slankis (1948, 1951) suggested that the fungus produces hormones that initiate short roots and regulate their development to nurture and maintain the association. Harley & Smith (1983) countered that the regulation was not complicated and the form of the ectomycorrhizal tip was primarily a result of physical factors. Regardless of the mechanism, the structure is clearly dictated by its being mycorrhizal.

In other types of mycorrhizae, changes in root structure with the fungal association are less apparent. However, in VA mycorrhizae, especially, the effects of mycorrhizae on root structure are still of major importance. Baylis (1975) hypothesized that VA mycorrhizae are of greater importance when root hair activity (in terms of number or length) is reduced. Thus, he argued that the symbionts comprising a VA mycorrhizae have coevolved based on root structure. St John (1980a) surveyed tropical trees and found, among his samples, that VA mycorrhizal activity correlated with reduced root hairs, supporting Baylis's hypothesis. Other studies, however, have demonstrated extensive and intensive VA mycorrhizal activity in grasses (e.g. Davidson & Christensen 1977; Caldwell *et al.*, 1985), plants with prolific development of root hairs. Despite some general evolutionary trends, VA mycorrhizae appear to have mixed effects on individual roots. About half of the papers that examine the influence of VA mycorrhizae on plant activity report increasing root/shoot ratios and another half report the opposite response (M. Allen, 1990). Hetrick *et al.* (1988) noted that VA mycorrhizae altered the root architecture of tallgrass prairie plants. In one axenic system, *Bouteloua gracilis*, a common grass on the North American Great Plains, root branching increased substantially with mycorrhizal formation when the phosphate source was predominantly Ca-phytate. When the phosphate source was NaH_2PO_4, no root responses were noted. These data suggest that the VA mycorrhizal effects on root structure are conditioned by the surrounding chemical environment. Thus, overall VA mycorrhizal effects on root structure probably result from a combination of the coevolutionary history of a plant species and the particular environment in which any given individual of the species resides.

Despite the obvious relevance of mycorrhizae to root–structure functioning, the importance of these associations is critical also to understanding the above-ground portion of the plant. After all, the plant photosynthesis, occurring primarily in the leaves, provides the carbon essential to both the roots and fungi. Several studies have shown that mycorrhizae not only increase growth, but also change the anatomy/morphology of the host stems and leaves. Daft & Okusanya (1973) noted an increase in number and size of vascular bundles in three of four crop species with mycorrhizae. In finger millet, increased lignification was demonstrated as well as a two-fold increase in vein diameter, 65% increase in leaf thickness, and 55% increase in mesophyll thickness (Krishna *et al.*, 1981). Conversely, no anatomical changes were observed in mycorrhizal versus nonmycorrhizal *Bouteloua gracilis*. Mycorrhizae also changed the gross morphology of some grasses. VA mycorrhizae changed the growth form of *Panicum coloratum* from prostrate to upright (Wallace, 1981), while in *Agropyron smithii*, the mycorrhizal plants were shorter in stature but had increased tillering (Miller *et al.*, 1987). VA mycorrhizal infection increased the leaf width on one variety tested (Di & Allen, 1990). *Leucina leucocephala* showed substantially altered morphology with mycorrhizae. Nonmycorrhizal plants exhibited continuously folded leaves to maintain water balance (this species exerts little stomatal control). VA mycorrhizal plants opened and folded leaves as appropriate to maintain CO_2 uptake and restrict water loss (Huang *et al.*, 1985).

Structure and functioning in the field

The dispersion in space of the external hyphal network, from the scale of an individual hyphal fragment to the distribution of mycorrhizae across a landscape, is poorly understood and of critical importance. It is the 'perception' and utilization of the surrounding environment that makes the relationship important to both organisms. The fungus reacts with only a small part of the soil whereas the relatively large size of the plant ensures that it will react with a large environmental unit. The fungus, because of its large surface-to-volume ratio, can intensively explore a soil volume. The plant explores a large volume less efficiently. However, a single plant may interact with several mycorrhizal fungi. The fungus may interact with several plants (see Chapter 6). In any case, understanding the scale of 'action' of the interacting individuals is important to describing the importance of mycorrhizae (Figure 2.7).

Horizontally, the structure of the association appears to be more a response to the environment than any developmental pattern beyond the mycelial structures discussed above. In undisturbed habitats, the mycorrhiza is patchy because of resource distribution associated with individual plants (e.g. Allen & MacMahon, 1985) and the distribution patterns of the host plants. For example, Chiarello *et al.* (1982) found that VA mycorrhizal hyphae connected adjacent mycotrophic plants but not nonmycotrophic plants in an annual community. They attributed a resulting complex ^{32}P allocation pattern within that community to these hyphal dispersion patterns. M. Allen *et al.* (1984a) found that the phenology of the fungus closely followed that of the host plant as co-occurring C_3 and C_4 plants had different VA mycorrhizal activity (e.g. in August, *Bouteloua gracilis*, the C_4 grass, had arbuscular activity and *Agropyron smithii*, the C_3 grass, had vesicles) with the same fungus. Anderson *et al.* (1983) reported that 10 cm^2 patches were most relevant to sampling VA mycorrhizal spore patches in a prairie whereas Allen & MacMahon (1985) found that even 4 cm^2 patches may not be small enough to delineate a patch in a semi-arid shrub steppe. Comparable work on other mycorrhizal systems has not been reported, but the interconnecting mycelium of the ascomycetous and basidiomycetous hyphae creates vastly different response surfaces (see Figure 2.4, and in Chapter 6).

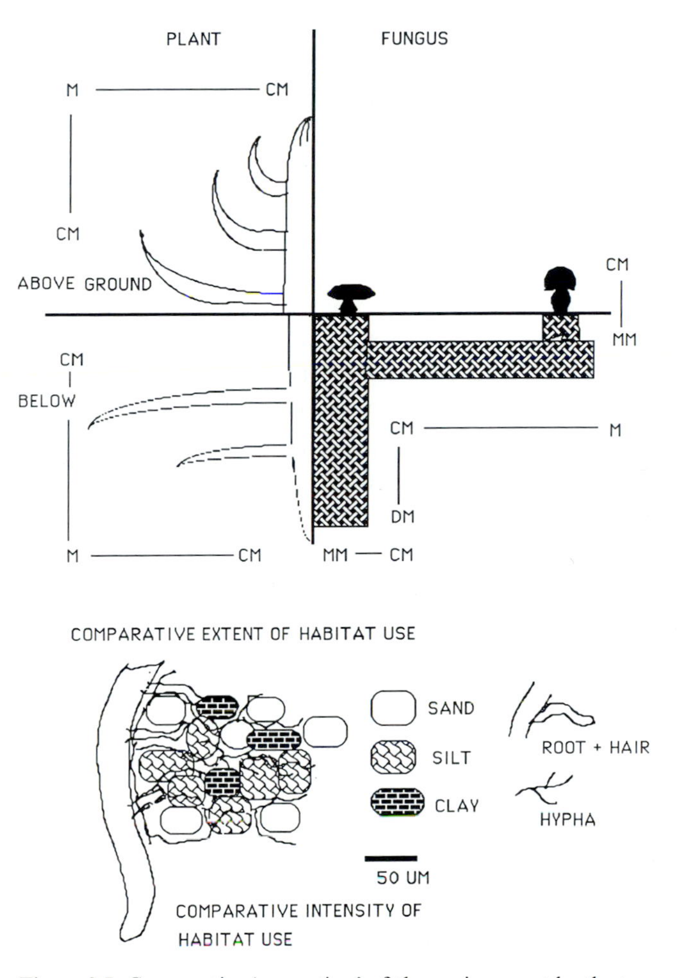

Figure 2.7. Comparative 'perception' of the environment by the two symbionts. The top portion describes the extent of habitat use. The plant utilizes a comparatively large volume of space. The fungus, because of its small size and its anastomosing mycelial network, utilizes a large area horizontally but that area is relatively thin. The fungus can extend deep into the soil but only along a root. Alternatively, the fungus utilizes a soil patch much more intensively (the lower portion of the figure). For example, in one cubic centimeter of soil, there may be only 2–4 cm of root, 1–2 cm of root hair, and 50 m of mycorrhizal fungal hyphae (e.g. see Allen, 1982). The thin diameter of the absorbing hyphae (1–2 μm) allows the fungus to penetrate through pores in silt particles, whereas even a fine root hair (> 20 μm) must move through large soil pores, between soil particles.

Animal activity appears to be a major regulator of the horizontal distribution of mycorrhizal activity. Even in 'undisturbed' ecosystems, localized disturbances create patches of low mycorrhizal activity. M. Allen *et al.* (1984b) reported that gopher activity reduced VA mycorrhizal spore density in a coniferous forest, and Koide & Mooney (1987) reported lower spore counts and inoculum density (as measured by bioassay) in gopher mounds than in serpentine grassland. Friese & Allen (1988) found that harvester ants concentrated inoculum as root fragments used to line tunnels and seed caches.

Disturbance can create distinct discontinuities in mycorrhizal inoculum or homogenize the mycorrhizal activity. On Mount St Helens, after the 1980 eruption, the burrowing activity of gophers created patches of high inoculum density in the matrix of sterile pumice (M. Allen *et al.*, 1984b; M. Allen, 1987a). Wind dispersal of inoculum created localized high inoculum patches of both ectomycorrhizal fungi on Mount St Helens (M. Allen, 1987a) and VA mycorrhizal fungi on a surface mine in the Great Basin (M. Allen, 1988b). Fire and smoke changed the spatial pattern of ectomycorrhizal activity in a Rocky Mountain coniferous forest (Harvey *et al.*, 1978). The responses of VA mycorrhizae are mixed, ranging from no reactions (Pendleton & Smith, 1983; Gibson & Hetrick, 1988) to reduced activity in patches with high fuel loads prior to burning (Klopatek *et al.*, 1988).

Several early reports suggested that mycorrhizal fungi were primarily distributed vertically near the soil surface where labile nutrients were being released (either by the mycorrhizal fungi themselves, e.g. Herrera *et al.* 1978, or from newly decomposing organic matter). Rommell (1938) used shallow trenches in a forest to detach the mycorrhizal fungi from the host and thereby separated mycorrhizal from saprophytic fungi. Sparling & Tinker (1975, 1978) found that most VA mycorrhizae in a Pennine grassland site were located in the surface 20 cm, a distribution pattern similar to that reported by Davidson & Christensen (1977) for the western North American Great Plains. Schwab & Reeves (1981) reported that inoculum density of VA mycorrhizal fungi was also highest in surface soils of the sagebrush steppes of Western North America.

Recent work, however, suggests that the vertical distribution is primarily a function of the distribution of resources and that VA mycorrhizal fungi can extend deep into the soil profile (see M. Allen 1988b). Zajicek *et al.* (1986) reported VA mycorrhizae at depths of 2.2 m deep on forbs in a tallgrass prairie. Virginia *et al.* (1986) found that VA mycorrhizae were concentrated up to 4 m deep in a sand desert in a nutrient-rich zone immediately above the water table. In western Wyoming, VA mycorrhizae extended downward in cracks in shale and sandstone parent materials to depths of 70 cm (M. Allen, 1988a).

The ability of an individual mycorrhiza to respond to resource patches in soil, e.g. a decomposing root fragment, has been hypothesized as an important component for mycorrhizal action (St John & Coleman, 1983). Mosse & Phillips (1971) and St John *et al.* (1983a), using sterile agar and soil, respectively, found that the hyphae were more extensive in the presence of organic phosphorus than of inorganic substrates. Read *et al.* (1985) described intensive mycorrhizal hyphal development associated with patches of decomposing organic matter. St John *et al.* (1983b) found that roots preferentially branched in litter bags over the surrounding soil matrix and suggested that this could indicate potentially increased mycorrhizal activity. M. Andrews, J. Richards and M. Allen (unpublished data) added P in discrete patches within the soil. No root response was observed for any of the four plant species observed. However, the proportion of the roots containing arbuscules, the VA mycorrhizal transport organs, increased dramatically in all actively growing species. These data suggest that the responses in space by the individual mycorrhiza are critical to mycorrhizal functioning.

Behavioral and structural/functional features associated with mycorrhizae, from the responses of individual hyphae to the responses expressed in whole plant growth form, affect the survival ability of plants during conditions of stress. The most important concept, as we attempt to understand the functioning of mycorrhizae, is that two structurally different organisms comprise the symbiosis and each utilizes the environment quite differently. The fungus is composed of an extremely fine hyphal matrix (with the absorbing hyphae generally substantially smaller in diameter than root hairs of grasses, 2–5 μm versus > 10–20 μm). This size allows the fungal symbiont to reach all but the finest soil pores. As mycorrhizal hyphal length can range up to 50 m or more per ml of soil (e.g. E. Allen & M. Allen, 1986), the capacity to exploit a given soil volume increases dramatically when a plant is mycorrhizal. In complement, the plant, by virtue of its direct carbon allocation, can send roots deep into the soil and widely across plant interspaces. Mycorrhizae cannot be perceived as a panacea for increasing plant production. However, the association appears to allow plants to acquire resources under times of acute stress, the 'ecological crunch' that Weins (1977) suggests is the crucial natural selection event for organisms (E. Allen & M. Allen, 1986).

3

Evolution

Mycorrhizae, as well as other mutualistic associations, are widespread. Nevertheless, mutualisms have been described as mathematically unstable ecological anomalies with limited long-term survival value. Mutualisms involving microorganisms and plants, especially, have been ignored in the coevolutionary literature (e.g. Boucher *et al.*, 1982). Most reviews rely either on fossil records that provide important morphological evidence but cannot delineate physiological interactions, or simply delineate those plant species that form mycorrhizae today. The development of viable models of the evolution of mutualism might aid in describing how, when and why mycorrhizae are important today.

The evolution of mycorrhizae may have been an essential precursor to the invasion of land by plants and their subsequent survival and expansion in terrestrial habitats (e.g. Bernard, 1909; Pirozynski & Malloch, 1975). The first land plants encountered a harsh environment. Soils contained no organic matter to hold nutrients and water, and true plant roots and root hairs had not yet evolved. On the positive side, the atmosphere contained high concentrations of CO_2 and light was not limiting to production. In an anthropogenic sense, these early plants had a choice: they could develop means of their own for gaining nutrients and water, or they could alter their relationship with the parasitic fungi that were already invading their rhizoids. The latter route was probably more feasible energetically, because the fungi already had a high surface-to-volume ratio, much higher than the plant was likely to develop. Also, a mutualistic relationship would reduce the negative effects of the parasite.

Although the evolutionary theory dealing with mycorrhizae concentrates on the reactions of the host, the fungi, as partners in the mutualism, were adjusting over time as well. The plant provides a ready source of simple sugars. Fungi that live saprophytically are forced to develop complex enzyme systems to break down complex carbohydrates that form the bulk of decomposable carbon. As virtually all mycorrhizal fungi approach obligate dependence on a host, a review of the mechanisms evolved by these fungi to 'steal' carbon from a reluctant host could help interpret the range of responses observed today among symbionts.

In this chapter, I will examine the hypotheses that: (1) mycorrhizae were important to the invasion of the land by plants, (2) the association was stable and important to the survival of host plants, and (3) through the long evolution of that mutualism, there was a genetic as well as an environmental basis for the mycorrhizal mutualism. While this discussion will be largely speculative, four lines of evidence contribute to our understanding of mycorrhiza evolution: (1) paleobiology and the evolution of fungi and plants; (2) extant plants and their mycorrhizal relationships; (3) the molecular biology of plant–fungal symbioses; and (4) models of mutualism and their possible applications to mycorrhizae. Based on these lines of evidence, hypotheses can be erected and falsified or validated to provide a better understanding of mycorrhizae and of mutualisms in general.

Paleobiology and the evolution of mycorrhizae

In 1904, Weiss described what appeared to be a mycorrhiza from 'lower' coal deposits. In what has become a classic contribution to understanding the evolutionary history of mycorrhizae, Kidstone & Lang (1921) presented a series of plates that show fungal structures closely resembling vesicles and spores in the rhizoidal cortex of *Rhynia* and *Asteroxylon* from the early Devonian. Several plates suggest that at least some of the structural elements of a VA mycorrhizal association were well developed by this time. These observations led the authors to postulate that *Rhynia* was probably mycorrhizal. But, as the authors carefully pointed out, structure does not necessarily imply that the same functional relationship existed then as now between host and endophyte.

Others found fossils that contained inclusions resembling VA mycorrhizae including the 'mycorhizome' of Andrews & Lenz (1943), the Devonian fossil *Psilophyton* (Stubblefield & Banks, 1983), and several Pennsylvanian coal fossil plants including lycopods, *Cordaites*, and *Psaronius*, a tree fern (Wagner & Taylor, 1981). Undoubtedly the most important recent discoveries are the arbuscules reported in *Antarcticycas* by Stubblefield *et al.* (1987a,b). In their second, more detailed set of micrographs, Stubblefield *et al.* (1987b) show a wide range of VA mycorrhizal structures in Triassic *Antarcticycas*. These fossils clearly demonstrate that the structural characteristics, at least, of the VA mycorrhiza were well developed by the Carboniferous.

Research on characteristics of the early terrestrial environment also contributed to the hypothesis that mycorrhizae were important to the invasion of land by plants. Jeffrey (1962) suggested that the development of a symbiosis (that led to archegoniate land plants) between an alga and an Oomycete like the development of the lichen symbiosis, was essential to land invasion. Nicolson (1975) and later Fitter (1985b) suggested that phosphorus limited plant growth in early terrestrial habitats. They concluded that since mycorrhizae are especially important in phosphorus uptake, the evolution of VA mycorrhizae between plant and fungus facilitated land plant survival. Early soils would have had little organic matter, and nutrient mineralization would have been slow because many of the most efficient saprophytes (e.g. ascomycetous fungi) had not yet evolved (Pirozynski, 1976).

Pirozynski & Malloch (1975) expanded the hypothesis for the development of the VA mycorrhiza between a primitive Oomycete and a chlorophytic alga. This relationship, they postulated, developed in the land-locked pools of the Silurian with the increasing aridity that occurred during the expansion of the continents. Pirozynski (1981) further suggested that higher plants evolved vascular systems as a result of the efficient transport by the fungal symbiont of water and nutrients from soil to plant.

The time and environment wherein mycorrhizae evolved obviously were contingent on where and when the host and fungus arose. Extant descendents of early land plants (e.g. bryophytes, *Psilotum*, lycopods) are mycotrophic. Discerning the ancestry of fungal partners is more problematical. Early students of mycorrhizal associations believed that the present-day VA mycorrhizal fungi were Oomycetes, known to have been present as early as the Silurian, when plants began to invade the land (Pirozynski, 1976). More recently, however, these fungi have been classified as Zygomycetes that appear to have arisen in the Carboniferous (e.g. Gerdemann & Trappe, 1974). Evidence for the classification of VA mycorrhizal fungi as Zygomycetes includes: the formation of suspensor cells comparable to those in Zygomycetes; the formation of zygospores in the genus *Endogone* (Gerdemann & Trappe, 1974); and chitinous (not cellulose) cell walls (Weijman & Meuzelaar, 1981). Alternatively, Jabaji-Hare (1988) suggested that chytrids also have chitin and that the fatty acid composition of *Glomus intraradices* and *vesiculiferum* and *Gigaspora*

Figure 3.1. Photographic representations of the evolution of types of mycorrhizae and their habitats, including: (A) a possible model of a Devonian landscape (Mount St Helens) that contained plants similar to early land flora such as mosses and algal mats, a mix of nonmycorrhizal and VA mycorrhizal plants; (B) a Carboniferous forest with tree ferns, horsetails, and club mosses, all VA mycorrhizal plants; (C) a mixed mycorrhizal Jurassic gymnosperm forest including VA mycorrhizal sequoia and ectomycorrhizal conifers; and (D) the highly diverse chaparral scrub lands with highly developed VA mycorrhizal *Adenostoma*, arbutoid mycorrhizal manzanita, and ectomycorrhizal oaks, all of relatively similar size and all with a high diversity of mycorrhizal fungi.

margarita resembles that seen in protostistan fungi more closely than that recorded for Zygomycetes. She suggests that a re-examination of the taxonomic status of VA mycorrhizal fungi is needed. One is tempted to ask if these fungi might be transitional between chytrids and Zygomycetes. The resolution of the classification controversy may have important implications for understanding when the symbiosis arose. Knowledge of developmental events and environmental conditions in which VA mycorrhizae developed should provide important insights into the current functioning of the symbiosis and into the regulation of community dynamics by mycorrhizae.

Tectonic activity in the late Mesozoic was associated with radical shifts in the evolution of plants and fungi. At this time, following the breakup of Gondwanaland, the Pinales in the Gymnospermae and the Basidiomycotina appeared. Ectotrophic mycorrhizae also first appeared (Pirozynski, 1981). By the early Cretaceous, an explosion had occurred in the diversity and predominance of angiosperms, roughly corresponding to the rise of the Ascomycotina. Pirozynski (1981) suggested that the development of forest mosaics dominated by 'pure' stands of trees with each root system encountering patchy soil conditions resulted in the development of low-diversity forests and high-diversity mycorrhizal communities. By the late Cretaceous, the wide array of mycorrhizal types found today had arisen (Figure 3.1), and the diversity of associated mycorrhizal fungal species has grown from a relative few in the early Mesozoic to literally thousands today. For example, Trappe (1977) estimated that up to 2000 species of mycorrhizal fungi were associated with Douglas fir alone!

Mycorrhizae and habitats of extant plants

Early work on mycorrhizal associations described groups of plants as forming specific mycorrhizal types and attempted to measure the reactions of different plant taxa to the differing associations. Stahl (1900) divided plants into nonmycotrophic (never forming mycorrhizae), facultatively mycotrophic (surviving with or without the symbiosis) and obligately mycotrophic (requiring mycorrhizae for survival). He then characterized plant families in terms of the prevalence of those mycotrophic groups. He suggested that the Chenopodiaceae and other related families were nonmycotrophic because he could not find mycorrhizae in the species that he examined. Grasses and other plants with fine roots tended toward facultative mycotrophy, and plants with coarse roots were classified as obligately mycotrophic. Of the primitive plants, he suggested that only *Equisetum* was nonmycotrophic.

As more surveys were completed, various workers have expanded Stahl's hypothesis (e.g. Gerdemann, 1968; Trappe, 1987). Their work, combined with the reports of Koske *et al.* (1985), indicate that all of the 'primitive' plants are VA mycorrhizal. Some of the more 'advanced' plant families developed mycorrhizal associations with more advanced fungi, such as ericaceous plants with ascomycetes and orchids with

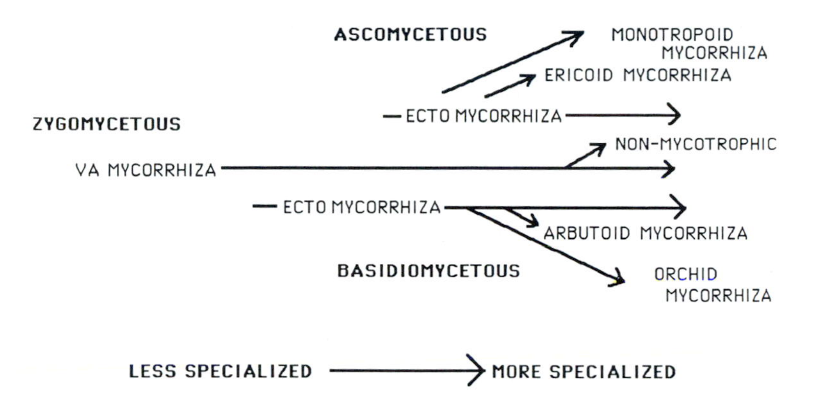

Figure 3.2. Evolution of specialization in mycorrhizae with increasing habitat specialization.

Rhizoctonia. Some, such as the Chenopodiaceae and Amaranthaceae, lost their mycorrhizal associations and became nonmycotrophic. These correlations support the idea that mycorrhizae evolved in early land plants and allowed them to expand across terrestrial landscapes.

Stebbins (1974) argued that, in plants, the most intense natural selection occurs in marginal habitats. As new habitats were being created by continental breakups and climate shifts, and as new plants evolved to occupy these habitats, the fungi presumably also evolved to adapt to the new plants and environments. The advanced fungi (Ascomycotina and Basidiomycotina) have both sexual and parasexual reproduction mechanisms allowing rapid genetic recombination, whereas the VA mycorrhizal fungi have infrequent anastomosis and no apparent sexual reproduction. Presumably the more advanced fungi were those best able to form a high diversity of mycorrhizal associations adapted to the new habitats. Figure 3.2 outlines a possible evolutionary scheme for the major types of mycorrhizae.

Nonmycotrophy is especially intriguing. Stahl (1900) alluded to the weedy nature of the nonmycotrophic plant families, especially the Chenopodiaceae and Brassicaceae. Nicolson (1960) suggested that lack of mycorrhizae may be a characteristic of early successional habitats; in particular, *Salsola kali*, an annual chenopod that is nonmycotrophic, was found as an early colonist in dune habitats. Reeves *et al.* (1979) and Miller (1979) noted that members of the Chenopodiaceae were frequent invaders of disturbed lands in semi-arid habitats, and reported a lowered inoculum in disturbed habitats compared with undisturbed habitats. Janos (1980) proposed that early successional plants were nonmycotrophic and that degree of mycotrophy increased with succession in the tropics. E. Allen & M. Allen (1980) found that both spore counts and VA mycorrhizal infection were reduced with disturbance. But, in early successional habitats, infection and spore density rapidly increased if mycotrophic plants were present. They also reported that the early successional annuals were predominantly nonmycotrophic. As these plants belong exclusively to advanced plant families, nonmycotrophy was hypothesized to be a derived character allowing a rapid exploitation of highly disturbed environments (e.g. Malloch *et al.*, 1980).

Newman & Reddell (1987) point out that differentiating mycotrophy at the family, or even generic, level is probably not acceptable. Thus, mycotrophy/nonmycotrophy is an expression of ecological adaptation and there may be no consistency within designated families or even genera. Berch & Kendrick (1982) reported that *Equisetum* was non-mycorrhizal, based on surveys in the boreal forests that often contain high water levels where *Equisetum* predominates. However, Lohman (1927) and Koske *et al.* (1985) found VA mycorrhizae in *Equisetum* when the plant was growing in dry soils. Numerous other species from purported nonmycotrophic families have been found to be mycorrhizal when growing in suitable edaphic conditions with adequate inoculum (e.g. *Carex*: Davidson & Christensen, 1977; Read & Haselwandter, 1981; E. Allen *et al.*, 1987; *Atriplex*: Williams & Aldon, 1976; M. Allen, 1983; *Polygonum*: Gerdemann & Trappe, 1974).

Careful examination of extant groups reveals a radiation in mycotrophy in increasingly more advanced plants. Nonmycotrophy appears to be restricted primarily to a limited suite of colonizing annuals in a few advanced families such as the Chenopodiaceae, Brassicaceae, Amaranthaceae, and Zygophyllaceae, and many of the hemiparasitic plants. The more complex types of mycorrhizal associations, such as ericoid mycorrhizae, orchid mycorrhizae, and monotropid mycorrhizae, are restricted to a limited number of highly advanced families. Thus, the extant plants and their mycorrhizal associations suggest that VA mycorrhizal associations are the conservative condition with other mycorrhizae and nonmycotrophy evolving later.

Another aid to understanding the development of mycorrhizae is to look for mycorrhizal associations in habitats resembling those where plants first invaded the land. Early primary successional sites have some of the important characteristics including low available nutrients and low organic matter. Sites with these characteristics include coastal sand dunes

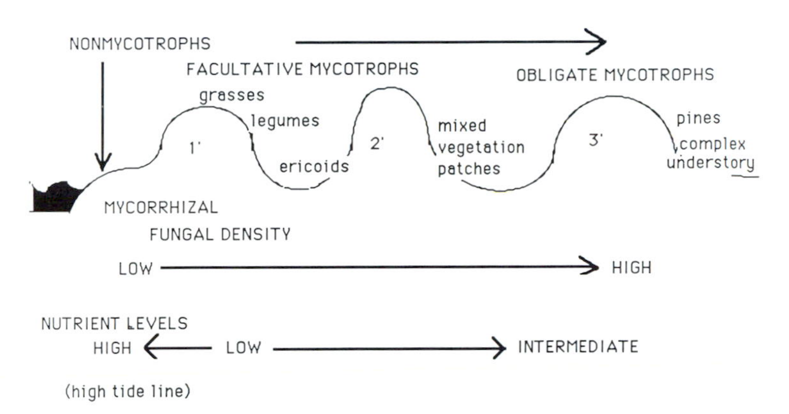

Figure 3.3 . Plant mycotrophy status in a gradient from the coast across island sand dunes. Also shown are the soil nutrient levels and mycorrhizal fungal density. Plants that do not form mycorrhizae are confined to the high tide lines that have high nutrient loading (from E. Allen & M. Allen, 1990).

and volcanos. At two such sites, the dominant pioneering plant species have been shown to be facultatively VA mycorrhizal (e.g. Koske & Polson, 1984; E. Allen & M. Allen, 1990). Nonmycotrophic plants do exist as infrequent adventives, but only in high nutrient patches such as the high tide lines where decomposing deposited materials provide both organic matter and released nutrients (Figure 3.3). On Mount St Helens, Washington State, USA, all colonizing plant species were examined for mycorrhizae and all were facultatively mycorrhizal (M. Allen, 1988b). Pioneer species on volcanic islands tend to be mycotrophic species (e.g. Simkin & Fiske, 1983; Henriksson & Henriksson, 1988). Studies on marsh vegetation might also provide some insights. Despite the common assumption that marsh plants are nonmycorrhizal, numerous papers report that when the soils become relatively dry, most of the plants are mycotrophic (e.g. Khan 1974; Read *et al.*, 1976; Mejstrik, 1984; E. Allen & Cunningham, 1983; Koh & Lee, 1984). Together, these data suggest that mycotrophy is a feature common to habitats that presumably most resemble early terrestrial environments.

Habitats conducive to most of the later evolving mycorrhizal types and to nonmycotrophic plants tend to be specialized and to have come about late in the earth's evolutionary history. The fungi comprising ericoid and orchid mycorrhizae, for example, are highly evolved organisms capable of extensive extraction of elemental nutrients from organic substrates, especially peats. The only plants consistently found to be nonmycotrophic are annual weeds adapted to invasion of disturbed, fertile habitats (E.

Table 3.1. *A list of some common weedy domesticated plants and their wild ancestors*[a] *that often show little or no response to mycorrhizae*

Crop	Ancestor
wheat	*Triticum*
oats	*Avena*
barley	*Hordeum*
sorghum	*Sorghum*
rice	*Oryza*
maize	*Teosinte* and a mix of weedy annuals
amaranth	*Amaranthus*
quinoa	*Chenopodium*
mustards	*Brassica oleracea*
macuasili	*Brassica campestris*
rochiwari	*Lepidium virginicum*

[a]From Harlan (1975); Bye (1979, 1985); Shuster & Bye (1983).

Allen & M. Allen, 1990). These weeds are especially able to invade and compete successfully with mycotrophic plants in highly fertilized agricultural or otherwise human-disturbed sites.

Agricultural breeding programs also provide some evidence that lowered dependence on mycorrhizae is a derived character. The plants most susceptible to domestication, and that make up many of our current dominant crops, are often weedy, nonmycotrophic or low-response mycorrhizal plants (Table 3.1). Varieties generally are selected based on their ability to produce high yields with artificial inputs of limiting resources (e.g. water, nitrogen). Moreover, limited experimentation on the response of host plants to mycorrhizae suggests that the more highly bred plants have a lesser response to the symbiosis (Azcon & Ocampo, 1981; Di & Allen, 1990).

Koide *et al.* (1988) reported that a cultivated annual (*Avena sativa*) had a greater response to mycorrhizae than did a weedy relative (*Avena fatua*) and concluded that cultivated plants have greater response than wild plants. However, a comparison with the weedy ancestor of oats was not conducted and weedy annual grasses often have less response to mycorrhizae than perennial grasses (Benjamin & Allen, 1987).

Agricultural selection methods might underline the notion that P fertilization alone can account for all mycorrhizal responses in a host (e.g. Graham *et al.*, 1987; Safir, 1987). Agricultural crops are selected for maximum yield dependent upon 'unlimited' water (through irrigation) and nutrients (through fertilization). Thus, a mycorrhiza, because of the carbon costs of supporting the fungus, can act as a parasite in a fertilized, irrigated cropland. The exception to easy artificial compensation is phosphorus. Phosphate is relatively immobile even with fertilization. In many croplands where all other limiting elements can move by mass flow, mycorrhizae greatly enhance phosphate uptake but increase uptake of other elements by lesser amounts. Thus, if phosphate is not limiting growth, the fungus would merely become a carbon drain.

In natural systems, the selecting agent is generally a stress imposed on survival to reproductive maturity of an individual and not *necessarily* the yield of a plant in a single growing season. By providing a limiting resource (e.g. ammonium) over a long period of improving stress tolerance during an 'ecological crunch' (e.g. drought: E. Allen & M. Allen, 1986), the mycorrhiza would improve the fitness of the plant despite some carbon loss. Thus, the range of selection forces in natural communities favors multiple-species infections and elicits a broad range of mycorrhizal responses.

Molecular biology and the evolution of mycorrhizae

Harley (1968) cautioned that, even though mycorrhizae and plant parasites are classified as distinctly different relationships, plant–fungal symbioses are actually gradients along a continuum of parasitism to mutualism. Some early workers entirely rejected the concept of mutualism. McDougall (1918) viewed mycorrhizae as reciprocal parasites, and McDougall & Liebtag (1928) concluded that a mycorrhiza was 'an antagonistic nutritive conjunctive symbiosis'. Curtis (1939) noted that, among his isolates, no apparent specificity existed between orchids and their fungal endophytes; because both the fungus and the orchid could be propagated independently, the mycorrhizal fungus was merely a harmless parasite on the orchid. Certainly, plant parasitic fungi have existed almost as long as their hosts. Did mutualism develop from parasitic relationships or are the two evolutionary lines fundamentally different? If so, can negative interactions between plant and mycorrhizal fungus occur, and how did the nonmycotrophic condition develop?

In any parasitic symbiosis, the host must survive to reproduction for the parasite to continue to survive. This has prompted the suggestion that in any parasitism, the parasite will become less virulent with time (e.g. Cook & Baker, 1983). Presumably any invasion of a plant by a fungus requiring carbon or other nutrients would initially have been a parasitism. Based on the assumption that the association would become less negative for the plant with time, what was initially a parasitism might have yielded a mycorrhizal association.

The recent theories of organism interaction at the molecular level might yield information on how mycorrhizae evolved and what current relationships may exist between plants and fungi. Specifically, did mycorrhizae result from some action by the plant that promoted invasion by specific fungi, or did the mutualism result from reduced resistance of the plant to these fungi as the relationship shifted from virulence to avirulence? Also, what are the mechanisms for interaction between the more advanced and often highly specific mycorrhizal types (e.g. orchid-*Rhizoctonia*, many ectomycorrhizae) and is there an interaction between mycorrhizal fungi and nonmycotrophic plants?

Woolhouse (1975) proposed an elaborate hypothesis to explain how endomycorrhizal fungi could invade a host. In this hypothesis, phosphate deficiency in a host cell results in the release of a signal, possibly in the form of phosphatases, that disrupts the action of protective host lectins. The fungi then invade the cell and the phosphate deficiency is corrected. Thereupon the lectins resume action, preventing pathogen invasion. This hypothesis stimulated interest in the mechanisms whereby mycorrhizae establish in phosphate-deficient plants, but not in phosphate-sufficient plants.

Subsequent studies have examined the role of external environmental characteristics that regulate mycorrhizal formation. These include postulates such as: phosphate deficiency induces increased membrane leakage which stimulates fungal invasion (Ratnayake *et al.*, 1978; Graham *et al.*, 1981); and volatiles secreted by plant roots direct fungal growth (Fries & Birraux, 1980; Koske, 1982; Gemma & Koske, 1988). Results from these studies support the idea that when the plant is under stress, it 'signals' the fungus to invade and 'correct' its deficiency.

Other efforts have concentrated on finding lectins that regulate compatability in more specific types of mycorrhizae. Bonfante-Fasolo & Perotto (1986) reported that different surface sugar residues were present in compatible and incompatible strains of ericoid mycorrhizal fungi. They suggested that recognition of these surface residues by lectins was involved in the early establishment of a mycorrhizal symbiosis.

Recent attempts to understand specificity in mycorrhizae have taken a different approach that may lead to important breakthroughs in understanding symbiont evolution. Molina & Trappe (1982a) carefully studied patterns of specificity among Pacific Northwest conifers and ectomycorrhizal fungi. They noted that fungi, forming highly specific host–fungus relationships, often invaded the cortex of non-host plants. Invasion resulted in a breakdown and lignification of cortical cells. Malajczuk *et al.* (1984) reported that invasion of eucalyptus by incompatible, conifer-specific mycorrhizal fungi stimulated high tannin production in the root cortex and lysis of both the fungal hyphae and the outer cortical cells of the host, whereas invasion by both broad-spectrum fungi and compatible fungi resulted in normal mycorrhizal development. These results suggest that compatibility might be more of a lack of recognition of the fungus by the host and that incompatibility is the active rejection of the fungus by the plant.

Observations on the interaction of VA mycorrhizal fungi and *Salsola kali*, a nonmycotrophic annual chenopod, supports the above hypothesis. Reduced growth of *S. kali* when mycorrhizal fungi were added (E. Allen & M. Allen, 1984) and reduced performance and survival in the field in inoculated compared with uninoculated plants (E. Allen & M. Allen, 1988) suggest an inhibitory interaction between plant and fungus. M. Allen *et al.* (1989a) recently found that arbuscules were also formed in *S. kali* but they lasted only for a day or two. Upon invasion, the plant immediately reacted to the fungus. Under fluorescence microscopy, the invaded cortical cells were observed to autofluoresce. Within 1–2 days, the root segment turned brown and frequently died. Simultaneously, the arbuscules disappeared. The external fungal hyphae angled off through the soil, frequently away from the root. In *Agropyron smithii*, a mycotrophic grass, such browning responses and root death were not observed. These responses suggest that nonmycotrophic plants recognize the invading fungus and reject it as a parasite. The mycotrophic plant fails to reject the fungus and forms a viable mycorrhiza. Others have substantiated the observations of arbuscule formation in clearly non-mycotrophic plants. For example, Tommerup (1984) found arbuscules and Glenn *et al.* (1985) reported VA mycorrhizal fungi in *Brassica napus*.

Anderson (1985, 1988) reviewed the concepts of specificity and recognition for mycorrhizae using existing models of elicitor action for plant pathogens. She suggested that both VA- and ectomycorrhizal fungi produced elicitors in legume cotyledon tissue. The role of suppressors is still under investigation (A. Anderson, personal communication). Nevertheless, a suite of observations and models suggests that the molecular relationship between host and symbiotic fungi (mutualistic or parasitic) may be fundamentally similar.

The elicitor/suppressor model suggests that the mycorrhizal fungus will invade any plant indiscriminately. The control of the fungal establishment

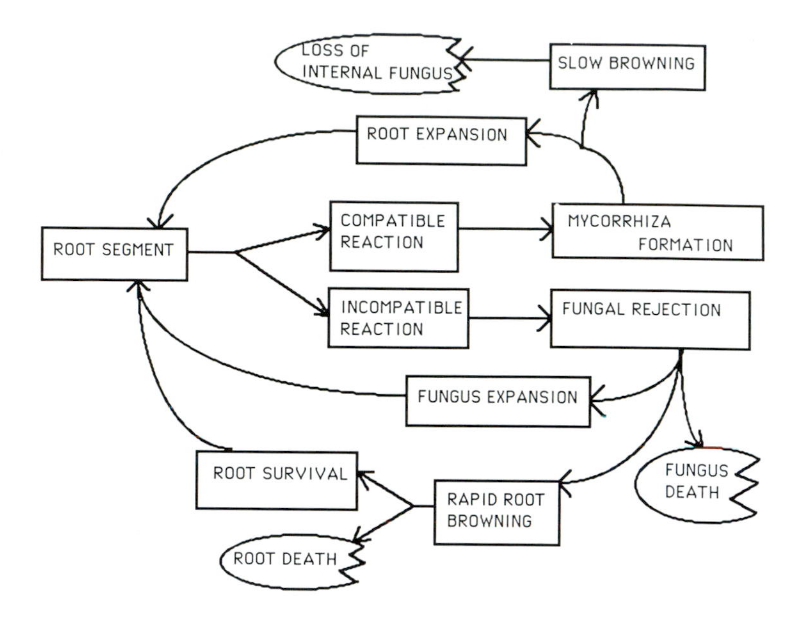

Figure 3.4. Compatibility reactions and the formation of mycorrhizae. If the host and fungus are compatible, mycorrhizal formation proceeds. If the host and fungus are incompatible, the host rejects the fungus by forming phenolics. If the fungus is able to gain carbon before rejection, it may expand and root death can occur (see M. Allen *et al.*, 1989a). If lignification is rapid, the plant can survive but the fungus may die.

is determined by the host rejection and continual re-invasion by the fungus (Figure 3.4). The system is never in equilibrium. For example, arbuscules of VA mycorrhizal fungi are digested rapidly, 1–2 days in *S. kali* (M. Allen *et al.*, 1989a) and 4–6 days in a compatible infection (Cox & Tinker, 1976).

Observations involving other mycorrhizal systems tend to support the elicitor/suppressor hypothesis. Rapid digestion of the mycorrhizal fungus by orchids is well documented (Harley & Smith, 1983). In fact, this digestion of the fungus appears to be the mechanism whereby the plant gains carbon from the fungus (Alexander & Hadley, 1985) and whereby the plant controls the *Rhizoctonia* fungus, often a virulent plant pathogen.

Evolutionarily, the elicitor/suppressor hypothesis may also explain the development of achlorophyllous plants (e.g. *Monotropa*) that derive their carbon and nutrients from the fungus and in which the fungus appears to gain nothing (Lewis, 1973). The fungus always invades an adjacent plant. If the plant can continually digest those invading structures without harming itself, and simultaneously gain carbon, so much the better for the plant. These responses indicate that greater effort is needed to understand the genetic as well as environmental bases for mycorrhizal formation.

Simulation models of mutualism and mycorrhizal relationships

Early community ecologists sought evidence for the existence of cooperative linkages among organisms that regulated important ecological processes. Clements (1916) proposed the 'super-organism' concept to explain the perceived stability of climax vegetation following succession. Allee *et al.* (1949) viewed communities as comprised of species that existed and interacted in a well-organized fashion. These ideas undoubtedly contributed to the perception that mycorrhizal fungi link the plants within a community and contribute to the well-being ('overyielding') of the plants within that community (e.g. Woods & Brock, 1964; Reid & Woods, 1969; Hirrel & Gerdemann, 1979; Janos, 1980, 1987).

Gleason (1926) argued that plants survived and reproduced as individuals, not as a structured community. Williams (1966) further asserted that individuals, not groups of organisms, are the unit of selection. By logical extension, then, any mutualism must be viewed as an anomaly. Wilson (1983) states:

> . . . the idea that soil organisms systematically modify their environment into a good growth medium for plants would be thoroughly rejected by most evolutionary ecologists (mycorrhizal fungi notwithstanding).

and:

> it can be shown theoretically that when the benefits of mutualism are shared among a group of neighbors, then the evolution of mutualism is impeded.

Theoretical modeling has indicated that mutualisms tend to be unstable because of two interacting factors. First, mathematical stability implies that neither partner can 'cheat', that is, take more than it gives. Second, there is an energetic cost to supporting a partner that takes away from the individual's energy allocation to its own gain, particularly in a competitive environment.

One might conclude that an impasse exists between the theoretical ecologists and mycorrhizal ecologists regarding the evolutionary significance of mycorrhizal mutualisms. The former suggest that mutualism is unstable and, therefore, cannot be widespread. The latter state that, not only is the mycorrhizal mutualism widespread, but because the mycorrhizal fungal hyphae connect individual plants, the symbiosis acts as an integrator of plant communities. Can these views be reconciled? Various data suggest that the two apparent contradictions are not totally in conflict.

Heithaus *et al.* (1980) proposed that a third species, a competitor or predator, can stabilize a mutualism by preventing the unlimited growth called for in the modified Lotka–Volterra competition models. Both competition between plants and predation of the mycorrhizal fungi by invertebrates and microbial parasites appear to be present in many ecosystems. E. Allen & M. Allen (1984) found that the grasses *Agropyron smithii* and *Bouteloua gracilis* responded to mycorrhizae with increased growth and water uptake significantly less when they were grown in monoculture than when they were grown in competition with the nonmycotrophic annual *Salsola kali*. The same competitive responses were observed in the field using the grasses *A. smithii* and *A. dasystachyum* (E. Allen & M. Allen, 1986). Daniels-Hetrick (1984) cited several examples demonstrating parasitism of the mycorrhizal fungus in the field by fungi and bacteria. Warnock *et al.* (1982) and Moore *et al.* (1986) found that invertebrate grazing restricted (or eliminated) mycorrhizal responses, and M. Allen *et al.* (1987) suggested that predation by nematodes or mites might have caused the observed reduced mycorrhizal fungal densities in years of high precipitation.

Dean (1983) contended that, if time lags are introduced into the growth of one of the mutualists, the mutualism can be perpetuated. Bethlenfalvay *et al.* (1982) found a lag period in seedling growth following invasion by mycorrhizal fungi. They postulated that this lag was caused by the energy diversion to the fungus. As the seedling grew and the roots began to deplete the available nutrient resources, the presence of the fungus became critical to obtaining the additional resources needed for continued growth. The production curve therefore resulted in a higher final mass in the mycorrhizal than in the nonmycorrhizal plant. Similar patterns have been reported by several researchers using a variety of test systems (Harley & Smith, 1983). This lag is the consequence of a temporary 'parasitism' of the plant by the fungus (e.g. Bethlenfalvay *et al.* 1982; Koide, 1985).

May (1981) and Dean (1983) also hypothesized that mutualisms are particularly sensitive to environmental perturbations and that this might explain the greater prevalence of obligate mutualisms in tropical as compared with temperate zones. Specifically, mutualisms consist of two interacting species. When either one is dramatically affected by environmental perturbation, both are subject to extinction. Although there are numerous examples of obligate mycorrhizal plants in temperate zone ecosystems, in temperate deserts, the most highly variable environment studied, facultatively mycotrophic rather than obligately mycotrophic plants dominate later successional ecosystems (E. Allen & M. Allen, 1990).

Despite the assertions of some modelers, the individualistic nature of each of the symbionts is rarely disregarded by mycorrhizal ecologists. Even when neighboring plants are infected by the same fungus, the phenology and different physiological responses of the plants involved result in different 'niche responses' (M. Allen *et al.*, 1984a) and different capacities for competition (Caldwell *et al.*, 1985; E. Allen & M. Allen, 1986, 1990). Read *et al.* (1985) suggest that one of the most important aspects of exchange of resources between plants via mycorrhizal hyphae appears to be in the ability of seedlings to utilize the carbon flowing from surrounding mature plants into the mycelial network.

Most models make two basic assumptions in suggesting that mutualisms (including mycorrhizae) are infrequent and unstable. The first is that equilibrium is essential to stability. Biological relationships are probably rarely in equilibrium. Mycorrhizal associations are characterized by continuous colonization, fungal establishment, host digestion or rejection, and recolonization by the fungus of newly expanded roots in new habitats; they are not characterized by an equilibrium in terms of biomass or 'numbers' of either symbiont. Secondly, the models assure that 'cheating' by either partner results in the loss of a viable mutualism. However, the sizes of the two partners comprising a mycorrhiza are radically different. To a plant, therefore, some 'cheating' by one fungus during a portion of the plant's life cycle would not catastrophically affect its overall fitness. Even if the mycorrhiza frequently reduces plant growth, if the alliance increased access to resources during 'crunch' periods, improved fitness would result (E. Allen & M. Allen, 1986). Thus, in a longer time scale, in which fitness becomes the defining parameter, the association can still remain mutualistic.

Most models assume that the limit to fitness is energy gain and that as the populations of the two symbionts increase, the greater demand on the host by the energy-dependent partner decreases the growth of the host (e.g. Mainero & del Rio, 1985; Vandermeer *et al.*, 1985). However, mycorrhizal fungi (and other mutualistic endophytes such as N-fixing bacteria), while requiring energy from the host, provide resources other than energy of value to the host. These resources are provided in greatest concentrations as populations of the host increase, with very little gain by the host at the low population levels. The models, such as those described in Boucher (1985), may be accurate only in weedy, fertile sites where energy gain may be the limit to fitness, or in wet areas where nutrients tend to be readily mobile and available. These habitats appear to be those where mycotrophy is selected against.

All of these exercises focus on the host as the organism of interest. The fungi forming mycorrhizal associations tended to evolve to utilize the simple sugars produced by the host and transported to its own root. This resource is more stable and requires considerably less energy to utilize than do the complex carbohydrates that represent the major detrital carbon inputs. Many mycorrhizal fungi have either lost or never had the enzyme capacities to assimilate complex carbohydrates and have become dependent on the host for survival. If the fungus cannot develop the characteristics necessary to maintain invasions and promote continued growth of the host, it will not survive. Thus, the representation of the fungal population, being obligately dependent on the plant population, is simply a +/0 relationship. This type of modeling approach is needed to understand mycorrhizal mutualisms.

General conclusion

In summary, we can speculate, both on the basis of the paleobiological evidence and on observations in extant, early successional habitats, that mycorrhizae may have been involved in the invasion of land and probably were important to the subsequent spread of plants across terrestrial habitats. Secondly, mycorrhizae form stable relationships and, with a greater attention to reality (especially regarding equilibrium and scale), these relationships are not inconsistent with individualism and, thus, with theoretical models. Finally, mycorrhizal associations have a genetic as well as an environmental basis. Extensive and intensive research are essential before these hypotheses can be satisfactorily addressed. As the organism is the basic selection unit, an organismal-level set of questions using information from lower hierarchical levels (e.g. cell–cell recognition phenomena) is probably essential before mycorrhizal systems can be fully utilized in both practical applications and as a basis for understanding population, community and ecosystem dynamics.

4

Physiological and population biology

In general, the study of mycorrhizae considers the effects of the mutualism on the physiology or production of individual plants. The two mutualists are rarely studied as individual organisms which interact to enhance the survival of each other. In part, the limited studies of the mutualism as a function of both organisms may be due to the differing scales necessary for study. Fungi, as with most microorganisms, are difficult to study because of their small size and our inability to separate species in the vegetative state. Also, both organisms comprising a mycorrhiza react to their common environment at different spatial and temporal scales (see Figure 2.7). For example, the mycorrhizal fungus predominates near the soil surface and often under the canopy, but the temperature and moisture regimes are quite different from those experienced by the leaves or even many of the roots. This microhabitat difference is generally ignored in theoretical models. In other types of mutualisms, such as bird pollinators and plants, the organisms react to similar microenvironments, e.g. air temperature and humidity, canopy structure, seasonal variation to light. In many lichen symbioses, the algal and fungal propagules escape the thallus as a unit whereas mycorrhizal fungus and plant reproduce entirely independently. Thus, when attempting to understand the processes that make mycorrhizal associations possible and successful, it is essential to understand the biology of each as independent organisms and then to derive hypotheses that test the selection factors that account for how and why these plants and fungi interact. In this chapter, I will describe how the symbiosis affects the physiology, growth, and reproduction of each of the mutualists, plant and fungus. As these are the parameters that contribute to fitness of an organism, I will attempt to generalize about the influence of the mutualism on fitness.

Genetic attributes of the symbionts

A difficulty in understanding the population biology of the organisms comprising a mycorrhiza is the problem of defining an individual versus a population of each organism. Harper (1977) describes different population levels for studying plants: plant parts as a population of units (e.g. leaves on a tree), clones of a genetically identical individual, and a collection of individuals. He equates vegetative reproduction with growth:

> . . . reproduction involves the formation of a new individual from single cell . . . growth . . . results from the development of organized meristems (p. 27).

Using this definition, loss of plant parts, limbs or 'individuals' in a clone, amounts to a reduction in mass. Others (e.g. White, 1984) also define fitness as numbers of progeny reaching maturity and note that the fitness of asexually reproduced individuals is often higher than in progeny resulting from sexual recombination. The central theme is that, although changes in biomass can be measured, measuring fitness of individuals is nearly impossible.

The same growth forms and definition problems in plant population biology also exist for fungi and colonial animals (e.g. Bell, 1984), only more so. Fungi contain several types of genetic information at different scales. Within a single cell, there are extrachromosomal genetic elements such as plasmids and viruses that act independently of, or in concert with, the nuclear or mitochondrial DNA. A single cell may also contain two genetically different nuclei, a heterokaryon, capable of independent multiplication and migration. Some cells can contain multiple nuclei. For example, a chlamydospore of a VA mycorrhizal fungus, a single-celled asexual spore derived from the tip of a single hypha, contains hundreds of nuclei (Cooke *et al.*, 1987) and sends out multiple germ tubes. It is unknown whether these nuclei are genetically identical or similar, i.e. whether the spore is homokaryotic or heterokaryotic. Coenocytic fungi (including the VA mycorrhizal fungi) have no regular cross-walls and even in fungi with septa (including ascomycetous and basidiomycetous mycorrhizal fungi), nuclei can and do migrate through a hypha. Hyphae, even from different 'individuals' of a species can fuse and exchange nuclear genes.

Rayner and colleagues (e.g. Rayner *et al.*, 1984) differentiate an 'individual' on the basis of somatic incompatibility, reactions that prevent hyphal fusion and genetic exchange between individual mycelial colonies. Fries (1987) found somatic incompatibility in the mycorrhizal fungus *Suillus luteus* and suggested that the same concept of an individual saprophytic fungus applies to mycorrhizal fungi as well. In addition, extranuclear genetic elements can migrate between incompatible fungi, although the extent to which exchange occurs is not known (e.g. Anagnostakis, 1987). When nuclei of different types occur in different cells, or different hyphae, sexual recombination can occur. Fungi may also undergo genetic recombination within a cell if the cell is heterokaryotic (2 different nuclei). This process, parasexuality, is frequent in many of the 'imperfect' fungi (Deuteromycetes) (e.g. Alexopolous & Mims, 1979). In the population biology of fungi, then, physically defining an individual is a major task. For purposes of this discussion, I use the following definition, recognizing the limits to these interpretations: an individual is a somatically compatible, attached mycelial network, and a population represents a group of individuals of the same species that are either not compatible or not physically attached and within a localized area.

Physiological ecology of the plant symbiont

In ecological time, the only important vocation of an organism is survival and production of offspring that will survive to maturity, or fitness. Production and fitness are not necessarily related. Many plants, such as those in arid habitats, do not have high productivity rates even under optimum growing conditions. Grime (1979), Chapin (1980), and Loehle (1988) have discussed the range of strategies plants can use to survive under different environmental conditions. Mycorrhizae alter a wide range of responses in the host (see Harley & Smith, 1983). Thus, in an ecological sense, improved acquisition of resources, that may not affect annual production but are important to survival only during key stress periods, may well determine the long-term survival of both symbionts.

Physiological changes in the host plant as a result of mycorrhizal infection have been extensively reviewed. Population ecology is, in part, the study of the survival of organisms that constitute a population, including those physiological properties that affect survival. I will discuss the physiological ecology of mycorrhizae, but only in the context of organismal survival. I refer the reader to Melin (1953), Marks & Kozlowski (1973), Harley & Smith (1983), and to the various mycorrhiza conference proceedings (e.g. Molina, 1985; Sylvia *et al.*, 1987) for detailed discussions of physiological interactions between mycorrhizal fungus and host. The following discussion emphasizes that the fitness of two complex groups of organisms which have interacted for at least 350 million years cannot be explained in terms of a single or even several simple parameters. My hope is that the reader will gain an appreciation of the wide range of interactions between the host and fungus that together have contributed to the evolution of the association through time.

A wide range of research efforts has demonstrated mycorrhizal responses in the host that affect host fitness. These responses result in changes in the realized or operational niche (*sensu* MacMahon *et al.*, 1981), changes in the actual niche, and alterations in a plant's reproductive ability. Mycorrhizae can alter the operational niche of a plant in two ways, by making previously unavailable resources available to the plant (adding axes in the niche hyperspace), or by increasing the potential gain of available resources (increasing an axis length or increasing the niche response) (Figure 4.1).

The ability of the mycorrhizal fungi to make unavailable resources available has sparked a continuing controversy (e.g. McDougall & Liebtag, 1928; Mosse *et al.*, 1981) but there are many examples. For instance, the transfer of organic carbon from soil organic matter to plant via the fungus by achlorophyllous plant mycorrhizae and orchid mycorrhizae (Harley & Smith, 1983) is a clear example of mycorrhizal provision of unavailable resources. In a monotropoid mycorrhiza, some evidence suggests that the fungus parasitizes a neighboring plant for carbon that is transported to the 'saprophytic plant' (Figure 4.2). In some orchid mycorrhizae (e.g., *Goodyera*), the fungus provides carbon to the non-photosynthetic protocorm and seedling until the photosynthetic machinery develops, whereafter the fungus apparently acts as other mycorrhizae, providing primarily inorganic resources (Figure 4.3).

Ericoid mycorrhizal fungi utilize complex organic N, an N source unavailable to the uninfected plant, and transfer some of that N as amino acids to the plant (e.g. Bajwa & Read, 1986). The ability of these fungi to provide that N is essential to the survival of the ericaceous plants, especially in heathlands, and explains, at least in part, the inability of other plants to survive in these habitats (Read, 1983).

More controversial are the hypotheses that VA and ectomycorrhizae have the capacity to gain mineral resources unavailable to the host. In organic soils, ectomycorrhizae may gain P or N from organic sources although the effectiveness of these fungi appears to be substantially less than that of ericoid mycorrhizae (e.g. Read, 1983). Sanders & Tinker (1971) concluded that VA mycorrhizal fungi utilize the same P forms that

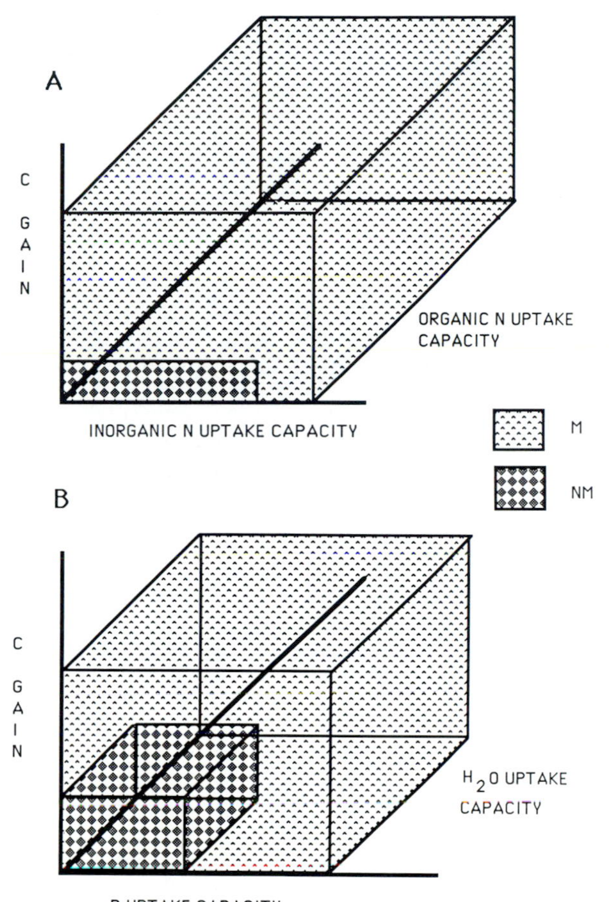

Figure 4.1. Mycorrhizae can alter the niche structure in two ways, by making resources available to the host that would not otherwise be available or by increasing the resource acquisition. (A) represents the case where mycorrhizae add an axis not present for a nonmycorrhizal plant. An example is the gain of organic N resources by ericoid mycorrhizae (from Bajwa & Read, 1986). (B) represents the increase in resource acquisition via mycorrhizal fungi from sources available to the nonmycorrhizal plant, but not necessarily in the same abundance, such as P or water (increased niche response from M. Allen *et al.*, 1984a).

plants do. Alternatively, mycorrhizal plant roots had external alkaline phosphatase activity whereas nonmycorrhizal plants had no measurable activity in sterile culture (M. Allen *et al.*, 1981a). The importance of this activity to the total P nutrition was not clear. Duce (1987) found no significant differences in phosphatase activity in rhizosphere soil of

Figure 4.2. The pathway of carbon flow to *Monotropa* or other achlorophyllous plants can be from several sources. Carbon may be fixed by the conifer, translocated to the roots, from the roots to mycorrhizal fungus, and thereby into the *Monotropa* plant where the plant digests the fungus (e.g. Bjorkman, 1960). Alternatively, the fungus breaks down complex organic matter from the soil and transfers it to the plant. Both pathways represent provision of a resource by the fungus that is not available to the plant alone.

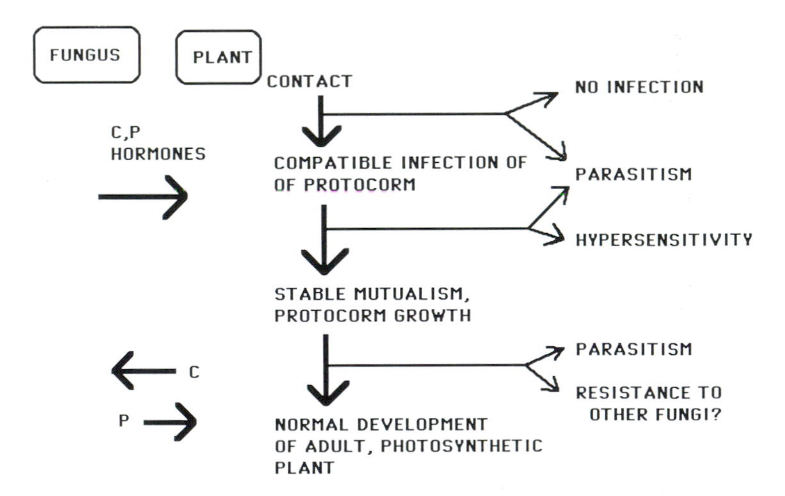

Figure 4.3. Life cycle and carbon and P nutrition of a photosynthetic orchid, *Goodyera* (derived from the work of Hadley and colleagues, e.g. Hadley, 1985; Alexander & Hadley, 1985; Alexander *et al.*, 1984). The directions of arrows represent the direction of resource movement, from fungus to plant or vice versa.

mycorrhizal and nonmycorrhizal plants. Interestingly, he showed that, as plants grew, the soil P pool that showed the greatest shift was in the unavailable inorganic P that was reduced by mycorrhizae. Few changes in the sizes of the other P pools were observed. Bolan *et al.* (1984) reported that P uptake could not be accounted for by depletions of soil solution and available P concentrations. Jurinak *et al.* (1986) proposed that VA mycorrhizal fungal hyphae produced oxalate crystals and, using a physical-chemical model, suggested that the oxalate crystals coupled with hyphal respiration could increase P solubilization and availability in clay soils. Both oxalates and soil CO_2 are affected by mycorrhizae. Oxalate crystals are common in some ectomycorrhizal fungal hyphae (Graustein *et al.*, 1977) and VA mycorrhizal fungal hyphae (Jurinak *et al.*, 1986). Pang & Paul (1980) and Kucey & Paul (1982) showed that a significant portion of the fixed carbon went to mycorrhizal fungal respiration. Knight *et al.* (1989) demonstrated that VA mycorrhizae increased soil CO_2 concentrations and that P availability in these soils supported the hypothesis of Jurinak *et al.* (1986). These data, although not complete, strongly suggest that VA and ectomycorrhizae, in addition to the ericoid, orchid and monotropoid mycorrhizae, have the capacity to gain soil resources that host plants cannot readily acquire.

Mycorrhizae certainly increase access to available soil resources, thereby increasing the niche response of a plant. This gain is due largely to the increased soil volume explored by the fungus. Hyphae extend outwardly from the root a few centimeters in VA mycorrhizae to, potentially, several meters in ectomycorrhizal systems. Direct transport of labeled P has been demonstrated frequently (e.g. Kramer & Wilbur, 1949; Hattingh *et al.*, 1973). VA mycorrhizal fungi transport P up to 7 cm distant (Rhodes & Gerdemann, 1975) and ectomycorrhizal fungi may transport specific ions up to several meters (e.g. Woods & Brock, 1964). Mycorrhizal fungal hyphae transport a wide range of minerals including N, Ca, Zn, S, P, K, etc. (see Melin, 1953; Gerdemann, 1974; Harley & Smith, 1983, for references to specific studies). More recently, substantial evidence has been presented for the transport of water from soil to host via the fungal hyphae (Duddridge *et al.*, 1980; M. Allen, 1982; Hardie, 1985; Bildusas *et al.*, 1986; Auge *et al.*, 1986).

The fungal hyphae, because of their small diameter (< 10 μm versus $>$ 20 μm for a root hair), also have access to small soil pores, especially in high clay or organic soils (see Figure 2.7). In some soils, this should increase the available resources. A limited amount of data also suggests that fungal hyphae may increase the efficiency of nutrient uptake as well as increase the surface area. For example, Cress *et al.* (1979) found that mycorrhizal tomato plants had a lower K_m than nonmycorrhizal plants, allowing for uptake of P at lower solution concentrations by mycorrhizal than nonmycorrhizal plants. Both possibilities need further study.

In some cases, mycorrhizae effect morphological responses that ultimately improve plant survival and growth. For example, Huang *et al.* (1985) demonstrated that VA mycorrhizae altered the leaf folding and orientation of *Leucaena leucocephala*. This leaf morphology regulated light interception and leaf energy balance thereby determined the plant growth rate. Krishna *et al.* (1981) found that mycorrhizae increased the numbers of vascular bundles in *Panicum millioides*. Slankis (1973) reviewed the extensive reports of morphological changes in the host plant that correlated with hormone responses, and M. Allen *et al.* (1980, 1982) reported hormonal changes in the host plant associated with mycorrhizal infection. These changes alter the ability of plants both to survive stress and to gain access to resources that ultimately improve the fitness of the plant (M. Allen, 1985).

Most mycorrhizal literature is concerned with above-ground plant production, especially of forests or crops. Within that context, the major research concerns are the resources provided by the mycorrhizal fungus that increase plant economic output, versus resources required by the fungus to support its own growth. For the mycorrhiza to be economically important, resources must be provided in levels beyond these which might be cheaply provided by other means.

Increased supplies of phosphate may result from mycorrhizal infection, a feature important to production agriculture. Agricultural mycorrhizal research also tends to look for a single 'mode of action', from the animal models widely used as analogs in plant physiology (e.g. phytohormones). This has led to assertions in the literature that mycorrhizae provide phosphate to the plant and that all other effects are artifacts resulting from the increased phosphate status (e.g. Safir, 1987; Graham, 1987). The ectomycorrhizal studies and more recent VA mycorrhizal publications have taken a broader approach. Extensive literature now shows a variety of resources provided by the fungus which may or may not be important in specific habitats, such as water, nitrogen, and micronutrients (see e.g. Melin, 1953; Reid *et al.*, 1983; Read, 1984; E. Allen & M. Allen, 1986, 1989; Trent *et al.*, 1989).

By increasing access to resources, mycorrhizae increase carbon fixation. This gain occurs primarily via increased photosynthetic rates. Levy & Krikun (1980) and Paul & Kucey (1981) demonstrated greater daily photosynthesis in mycorrhizal than nonmycorrhizal plants using ^{14}C absorption methods. M. Allen *et al.* (1981b, 1984a), Trent *et al.* (1989), and Reid *et al.* (1983) using CO_2 flux measurements demonstrated greater instantaneous photosynthesis in mycorrhizal than in nonmycorrhizal plants. Mycorrhizal plants can also take up more carbon in drought cycles than can nonmycorrhizal plants. Mycorrhizal *Bouteloua gracilis* maintained more open stomates at lower soil water potentials than did nonmycorrhizal plants (M. Allen *et al.*, 1981b, 1984a), and wheat plants infected by *Glomus fasciculatum* had stomates open for up to 3 days longer during drought than nonmycorrhizal wheat plants (Allen & Boosalis, 1983). In a recent field study, Trent *et al.* (1989) demonstrated greater photosynthesis and stomatal conductance while the plants maintained equal to lower leaf water stress during grain filling of mycorrhizal versus nonmycorrhizal winter wheat with no significant nutritional differences.

Mycorrhizae can also increase total seasonal carbon gain. Flower phenology was delayed with drought when plants were mycorrhizal versus when nonmycorrhizal (E. Allen & M. Allen, 1986) and leaf mortality was reduced with mycorrhizal infection (E. Allen and M. Allen, unpublished data). These experiments were performed in habitats where drought restricts the length of the growing season. The retention of actively growing leaves later into the drought period provides greater seasonal carbon gain, and perhaps the chance to remain active until the next precipitation event. Other reports also note improved leaf retention (e.g. Kormanik, 1985) even without drought stress. Especially in perennial plants, the long-term carbon gain, whether allocated for reproduction or growth, can be important to the survival and fitness of the host plant.

Often, field studies on mycorrhizae fail to validate the dramatic results observed in glasshouse studies (Fitter, 1985b). This has led many ecologists to suggest that mycorrhizae are not important in the field and can be safely ignored. Alternatively, as suggested by E. Allen & M. Allen (1986), it may be that mycorrhizae have advantageous effects on plants only during times of stress. In their study, effects of mycorrhizae on stomatal resistance were observed only during drought. E. Allen & M. Allen (1986) propose that often the critical mycorrhizal response is during these 'ecological crunches' (from Weins, 1977) and that these events represent the agents of selection during which mycorrhizae are particularly important to the plant. Read *et al.* (1985) suggested that seedlings tap into existing mycorrhizal mycelial networks that provide both nutrients and carbon to the establishing plant. This process could contribute to enhanced survival of that plant and, thus, its fitness. This shift in the

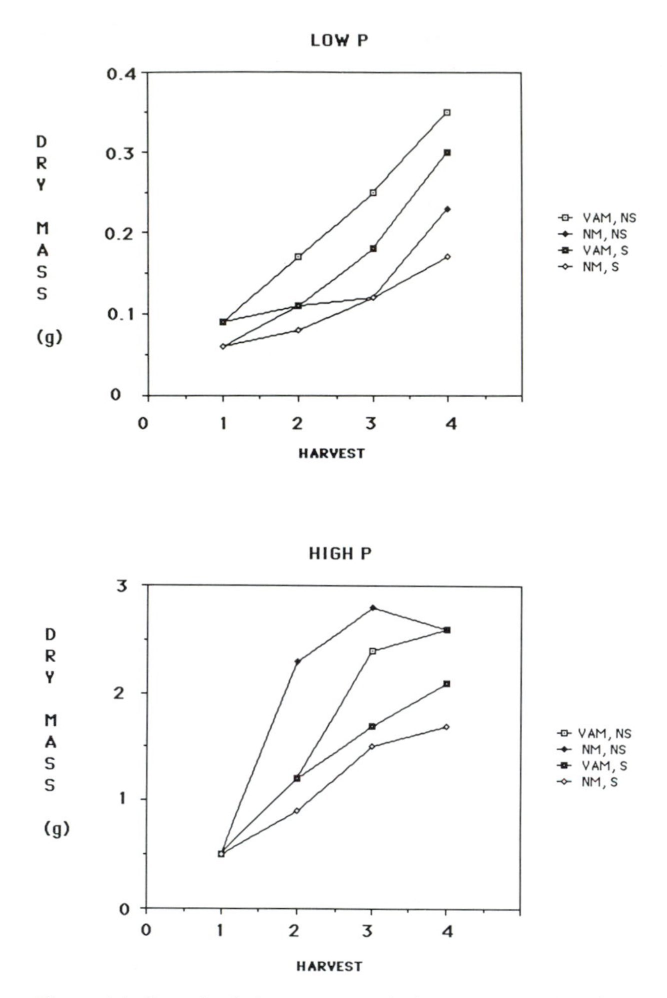

Figure 4.4. Growth of *Agropyron smithii* in response to P and water stress in sequential harvests. Upper graph represents above-ground plant mass when the grasses were grown in high P soil and lower graph represents growth in low P soil. Shown are monthly harvests of mycorrhizal with no water stress (⊡), nonmycorrhizal with no water stress (◆), mycorrhizal with water stress (■) and nonmycorrhizal with water stress (◇), from Duce (1987).

'instantaneous niche' (*sensu* MacMahon *et al.*, 1981) may well represent the time scale in which mycorrhizal associations need to be studied.

Mycorrhizae are also primarily studied at the scale of the whole host plant, not at the scale of infected roots. M. Allen *et al.* (1989c) found large plant-to-plant and year-to-year variation in fungal infections and sporulation in the field despite careful experimental control and suggested that the scale of the fungal action may be at the level of the individual root. M. Andrews *et al.* (unpublished data) noted that, in the presence of small patches of P addition, mycorrhizal activity increased significantly, but the response could not be detected at the whole plant level. These small-scale responses may be the major mycorrhizal action that, in conjunction with the short-term crunches, affect plant fitness, particularly of long-lived plants in the field.

Although mycorrhizal infections are normally considered as beneficial to plants, the reaction of the plant can range from positive to negative depending on the genetic and environmental parameters (see Chapter 3). Mycorrhizal fungi represent a carbon drain on a plant, especially at the seedling stage. Buwalda & Goh (1982), Bethlenfalvay *et al.* (1982) and others have demonstrated a lag in seedling growth that they suggest represents the carbon lost to the fungus in a growing seedling. Duce (1987) found that, under less stressful soil P conditions, VA mycorrhizae initially reduced above-ground phytomass, but that when the plants were P-stressed from the beginning, mycorrhizae increased plant growth even as seedlings (Figure 4.4). Some cases of inhibition of plant growth by mycorrhizae have been reported. Allen & Boosalis (1983) noted that VA mycorrhizal infection by *Glomus mosseae* led to increased stomatal opening with drought to the point that the wheat seedlings wilted, although *G. fasciculatum* improved plant drought tolerance (figure 4.5). Modjo & Hendrix (1986) reported increased growth of tobacco plants when the mycorrhizae were eliminated, although they did not eliminate the possibility of factors such as nutrient release from dead micro-organisms or reduced pathogen activity as causal agents.

Many of these observed responses may be artifacts of the cultural conditions. Read *et al.* (1985) note that in many natural ecosystems, the mycorrhizal fungus exists as a mycelial matrix tying together many plants. They suggest that a seedling actually taps the carbon and nutrient resources of that entire matrix, resulting in improved benefits for the seedling as opposed to the one-on-one interaction created by the pot studies described above. This could have profound implications for seedling survival depending on whether the plant was establishing in a seasonal, disturbance regime (created by soil disruption or by climatic disruption) or whether seedlings were establishing in an already established mycelium. These alternative scenarios need further testing.

As described in Chapter 3, gradients in plant mycotrophic status have been shown by a number of researchers (e.g. Stahl, 1900). If genetic incompatibility between host and fungus exists, the fungus can reduce the

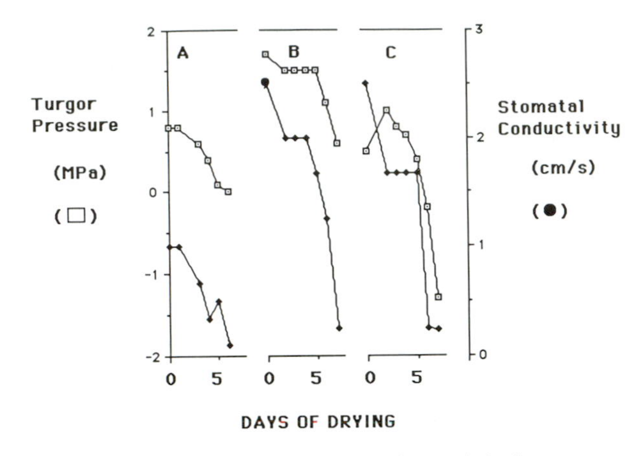

Figure 4.5. Responses of wheat to a drying period when nonmycorrhizal (A) or when mycorrhizal with the fungus *Glomus fasciculatum* (B) or *Glomus mosseae* (C). Shown is the mid-day stomatal conductance (◆), and the leaf turgor pressure (⊡), from Allen and Boosalis (1983).

fitness of the plant associate. E. Allen & M. Allen (1984, 1988) reported reduced growth and seedling survival of nonmycotrophic annuals with the addition of mycorrhizal fungi. The root browning of incompatible host–fungus interfaces (see Chapter 3) cannot aid the growth and survival of these seedlings (e.g. Molina & Trappe, 1982b). Additional studies on the genetics of compatibility and host fitness and distribution are needed.

Often, the effects of mycorrhizae on a host are described in the context of one fungus–one host, based on the results of pot studies. Nevertheless, a wide diversity of mycorrhizal fungi can exist even with a single host. Trappe (1977) noted that up to 2000 species of fungi could associate with a single host tree and he suggested that multiple fungi invading a single plant confer advantages beyond adding greater surface area for nutrient absorption. Considering that each species of mycorrhizal fungus can acquire different resources, Trappe's hypothesis is reasonable and should be tested further. For example, the ectomycorrhizal fungus, *Hysterangium*, grows in a concentrated mat and produces high concentrations of oxalates that radically alter the availability of cations (Cromack, 1981) and, by binding those competing cations, increase the availability of P (Jurinak *et al.*, 1986). Other fungi, such as *Cenococcum*, that will associate with the same root, are diffuse in the soil but can tolerate low soil water and increase water uptake in the host during drought (Mexal & Reid, 1973). Different VA mycorrhizal fungal species have different external hyphal characteristics (e.g. Abbot & Robson, 1985) that can allow different resource acquisition. These fungi can simultaneously occupy the same root system and can differently affect the root segments that they occupy.

Cluett & Boucher (1983) suggested that two mutualisms, nodulating N-fixing bacteria and mycorrhizae, have only indirect effects on each other. That is, they affect each other only through the resources that each supplies independently to the host. The mycorrhizae provide soil minerals and the bacteria provide fixed N, each of which increases plant growth independently. This increased carbon allocation to the roots presumably also supports the other endophyte. This hypothesis should be further evaluated since there are alternative hypotheses concerning the interactions of root inhabitants (e.g. Mosse *et al.*, 1981). The potential for competition for invasion sites between mutualists, between mycorrhizal fungi and pathogenic fungi, and between root symbionts that are completely different (e.g. fungi and bacteria) is important in determining plant growth and survival. The competition between different strains of *Rhizobium* for invasion sites is well known, especially as both effective and ineffective strains can occupy the same sites with considerable effects on plant growth.

The results of studies of relationships between mycorrhizae and other mutualists, such as N-fixers, are not definitive. Although many studies show only additive effects (Cluett & Boucher, 1983), many others do not. Neither *Rhizobium* nor mycorrhizal inoculation altered the growth and reproduction of *Hedysarum* in the field, but together they both contributed N to the plant and increased growth and reproduction (Carpenter & Allen, 1988). Results with agricultural crops are conflicting (e.g. Bagyaraj, 1984). There is probably little competition for infection sites between *Rhizobium* and VA mycorrhizal fungi as rhizobia invade root hair cells, rare for VA mycorrhizal fungi. Because ectomycorrhizae suppress root hair formation (Harley & Smith, 1983), these associations may reduce fixation, but this possibility is untested. Rose & Youngberg (1981) suggested that the interaction between *Frankia* and VA mycorrhizae was synergistic in *Ceanothus velutinus* as both growth and nutrition concentrations in the dual-infected plants were much greater than the responses when each endophyte was added separately. Comparable studies have not been completed with ectomycorrhizae and *Frankia* associations. Clearly, more efforts are needed. Evidence based on sites of invasion and molecular responses to invasion, not just plant growth, will be required to evaluate the potential for synergism among the differing interactions.

Marx (e.g. 1969a,b) found that ectomycorrhizae reduced the establishment ability of parasitic fungi in trees. In the ectomycorrhizal systems the mechanisms were apparent; the fungus surrounded susceptible root tissue and physically prevented detrimental organisms from invading. Kowalski (1980, 1982) found that the saprophytic fungi associated with mycorrhizal trees of *Pinus sylvestris* inhibited the growth of *Cylindrocarpon destructans*, a parasite of the tree. This inhibition then determined whether or not the replanted stand of trees survived. These data clearly show that some ectomycorrhizae have the ability to reduce disease and enhance plant fitness.

In the case of VA mycorrhizae the evidence is not nearly as clear. These fungi do not cover the root as do ectomycorrhizae, nor do they appear to produce antibiotics. Several reports show a wide range of interactions between mycorrhizal fungi and parasitic microbes (see Schenck, 1981; Bagyaraj, 1984). Most studies showing reduced pathogen activity with VA mycorrhizae attribute that resistance to improved P nutrition (e.g. Davies & Menge, 1981; Graham *et al.*, 1982), although other studies suggest that fertilization with a wide range of nutrients can produce the same response (e.g. Reis *et al.*, 1982).

Fitness of the plant symbiont

Organisms live, reproduce, and die as individuals. Genetic alterations occur within an individual (mutation) and between individuals as different individuals fuse, undergo recombination and separate. Selection then acts on these individuals leaving the most 'fit' to survive and reproduce. Ideally, then, the study of the individual organism and the environmental factors that regulate its interactions with other organisms would best allow us to understand mycorrhizal functioning. An organismal approach, outlined in detail by MacMahon *et al.* (1981), can be used as a heuristic tool for studying the formation of mycorrhizal associations and for outlining the importance of the mycorrhizal symbiosis for the survival of each partner. This perspective is especially important in mycorrhizal mutualisms (compared with many other mutualisms) in that each fungus forms associations with multiple plants and each plant forms associations with multiple fungi.

Each organism has a distinct set of genes that determines its range of growth and developmental possibilities, its fundamental or 'hereditarily determined' niche (MacMahon *et al.*, 1981). Recall that niche is a function of the organism and there are no 'empty niches', only unused resources (e.g. Whittaker *et al.*, 1973). Because the potential for forming a mycorrhiza is genetically determined (see Chapter 3), then the symbiosis

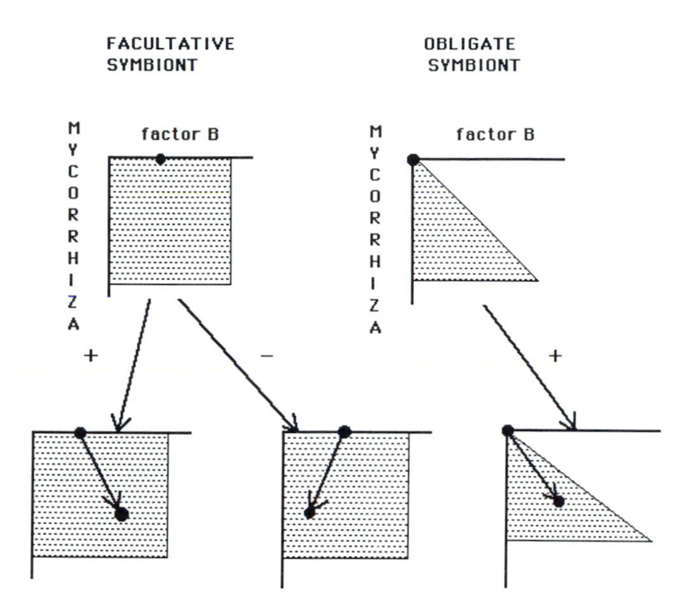

Figure 4.6. Effects of mycorrhizae on actual niche for the association to be important to the host plant, be that plant facultatively or obligately mycorrhizal. (●) represents the actual (or instantaneous niche) and (▨) represents the potential (or hereditary niche) (terminology derived from MacMahon *et al.*, 1981). Factor B is a resource. The arrows represent the shift in the actual niche that occurs when infection goes from 0 to some value *n*. In a facultative symbiont, the addition of mycorrhizae can either increase (+) the acquisition of a limiting resource (e.g. P) or decrease (−) it (e.g. carbon gain under low light). In an obligate symbiont, the plant cannot gain resource B without a mycorrhiza, therefore the addition of the mycorrhiza always improves (+) the acquisition of resource B .

is one axis of the potential niche of each of the symbionts. The environment constrains the hereditary niche, reducing it to a single, instantaneous position in the hyperspace, the actual niche (MacMahon *et al.*, 1981). If mycorrhizae are important to an organism, then the actual niche may be at a different point along a second or third or other axis. This concept is especially critical in the context of ecological crunches (e.g. E. Allen & M. Allen, 1986). The mycorrhizal and nonmycorrhizal plants may have little difference in the hereditary niche. However, when the plant is provided with its symbiont, the instantaneous position may be at a location that provides it with the ability to survive that crunch. An example is the ability of the mycorrhiza to provide increased water uptake during a short drought (E. Allen & M. Allen, 1986). The ability of an organism to tolerate a broad range of selection forces dictates its survival, and a symbiotic association that changes that tolerance is critical to survival. Therefore, a mycorrhiza is important only if it changes the actual niche in a critical direction (Figure 4.6).

The response of an organism to the nth number of axes that define the niche of that organism is a function of its physiology. The study of ecology, in part, began as a correlation between the distribution of organisms and the physiological traits that allow them to live in differing environments (McIntosh, 1985). Each organism responds to its environmental cues, biological, chemical or physical, and lives and reproduces or dies as its tolerance range is tested. The difficulty in assessing the importance of any parameter, be it a mycorrhizal relationship or water availability, is that an individual can only rarely be studied for its range of potential responses, only a population can be assessed. Fitness, or the ability of an organism to produce surviving offspring, is a function of an individual, not the population. Population ecology, or the study of the survival and reproduction of genetically related individuals, is the basic unit of study required to understand the importance of a relationship such as a mycorrhiza.

Most of the literature on the study of mycorrhizae addresses the effects of mycorrhizae on plant growth. Attempts to establish plants in disturbed habitats or in previously unoccupied habitats have demonstrated the importance of mycorrhizae to plant survival. The establishment of trees with ectomycorrhizal associations in such areas as the Ukrainian steppe (e.g. Schemakhanova, 1962), the Caribbean islands (e.g. Briscoe, 1959), and the American Great Plains (Goss, 1960), and the importance of VA mycorrhizae to plant survival in mined or desertified habitats (e.g. Aldon, 1975; M. Allen, 1988a), demonstrates the importance of mycorrhizae to plant survival.

The effect of any factor on plant fitness is related to the reproductive output of that plant (Harper, 1977). Because mycorrhizae may improve plant production or plant physiological status, it is often assumed that the association also improves reproduction or seed production of the host. However, few studies have tested this assumption. In VA mycorrhizae, results from the few agricultural studies are frequently contradictory (e.g. Powell, 1984; Hayman, 1987), due to difficulties in obtaining adequate conditions for the plant to complete its life cycle in small pots in a glasshouse, and to eliminate infection for nonmycorrhizal treatments in the field. Khan and colleagues (e.g. Khan, 1972, 1975; Saif & Khan, 1977) and others more recently (see Powell, 1984 for references) suggested that VA mycorrhizae increased grain production in low-fertility and low-inoculum fields only. Carpenter & Allen (1988) found that the inoculation of both VA mycorrhizae and *Rhizobium* enhanced flower and fruit production in the legume *Hedysarum boreale* but that when each endophyte was added alone, it did not significantly affect reproduction. The study site soils were highly fertile. Yocum (1985) found that mycorrhizae enhanced reproductive output in mycotrophic plants in low P soils and Yocum *et al.* (1987) also reported enhanced grain yield in winter wheat with the addition of VA mycorrhizae. Koide *et al.* (1988) found increased seed production in two annual grass species. Trent *et al.* (1989) reported greater grain filling in mycorrhizal than in nonmycorrhizal wheat. No studies on the effects of ectomycorrhizae on reproduction appear to have been published, perhaps because of the long time required to assess reproduction in long-lived plant species that are commonly ectomycorrhizal.

Orchid mycorrhizae represent special cases affecting the role of the fungus in 'fitness'. The seeds will not, in nature, germinate without the fungus and the protocorm often will not develop without the fungus (Hadley, 1985). The non-photosynthetic portion of the life cycle of an orchid apparently requires fungal feeding for growth, although in the photosynthetic portion of its life cycle the mycorrhizae may primarily provide mineral nutrients (Figure 4.3).

Mycorrhizae may also affect ramet development, although this may or may not be considered reproduction (*sensu* Harper, 1977). Miller *et al.* (1987) reported increased tillering in *Agropyron smithii* with VA mycorrhizae. Many mycorrhizal plants, e.g. aspens and many grasses, expand primarily by ramet growth, rarely by seeds. Mycorrhizae often enhance the growth of these plants and might be presumed to provide resources that will aid in the energy available for asexual expansion or reproduction.

To summarize, mycorrhizae probably enhance the fitness of most plant species but not in all environmental situations. Nevertheless, this conclusion is based on indirect evidence and there is an essential need for further studies that will give stronger evidence for the roles of mycorrhizae in plant fitness in a wide range of environmental situations.

Physiological ecology of the fungal symbiont

Describing the role of the mycorrhizal association in the fitness of the fungi is more difficult than describing the host plant's fitness because an individual cannot be distinguished except with the utmost difficulty for experimental study. Because of this, the interactions are best described based on the resource needs of the fungus that are provided by the host and the ability of the fungus to expand its range and, presumably, increase its fitness.

The mycorrhizal fungus is obligately dependent on a host plant for its energy and presumably for other resources as well. There is a range in the specificity of the fungus to its host; some, such as *Alpova diplophloeus* on alder, are highly specific (Molina & Trappe, 1982a), whereas many species of *Glomus* can form a viable symbiosis with virtually any VA mycorrhizal plant (e.g. Bowen, 1987). The 'natural history' and resource limits to growth in the field ultimately define the characteristics that limit the fitness of an organism.

Many 'higher' mycorrhizal fungi can be grown in pure culture with appropriate nutrients and carbon sources (see Schenck, 1982). The mycorrhizal fungi of 'saprophytic' (achlorophyllous) plants and of orchids provide carbon to the plant, apparently via host digestion of the internal organs of the fungus (e.g. Lewis, 1973; Hadley, 1984). Ericoid mycorrhizal fungi break down and utilize organic N from complex carbon molecules (Bajwa & Read, 1986). Some ectomycorrhizal fungi also have the enzymes necessary to utilize complex carbohydrates (Trojanowski *et al.*, 1984). Norkrans (1950) studied a gradient of *Tricholoma* species from saprophytic to mycorrhizal. The ability to break down cellulose was negatively associated with the ability to form mycorrhizae.

Despite these data, the generalization that most energy for fungal growth is from the host still stands. The excellent early work of Harley and colleagues (see Harley, 1969; Harley & Smith, 1983, for citations and reviews) clearly demonstrated that carbon in the form of simple sugars was transported from plant to fungus, whereupon the fungus converted those sugars to trehalose and other complex organic molecules which effectively prevent reabsorption by the host. Bjorkman (1960) found that *Monotropa* absorbed ^{14}C from nearby labeled conifer trees, suggesting that even these achlorophyllous plants derived their carbon from the mycorrhizal fungus that derived it from the neighboring trees. Castellano & Trappe (1985) found that many of the mycorrhizal fungi associated with *Monotropa* were common ectomycorrhizal fungi of nearby conifers. Lewis (1973) notes that fungi, in general, are structured to acquire but not give up carbon.

VA mycorrhizal fungi have not been grown in pure culture, suggesting obligate dependence on the host. Ho & Trappe (1973) found that spores

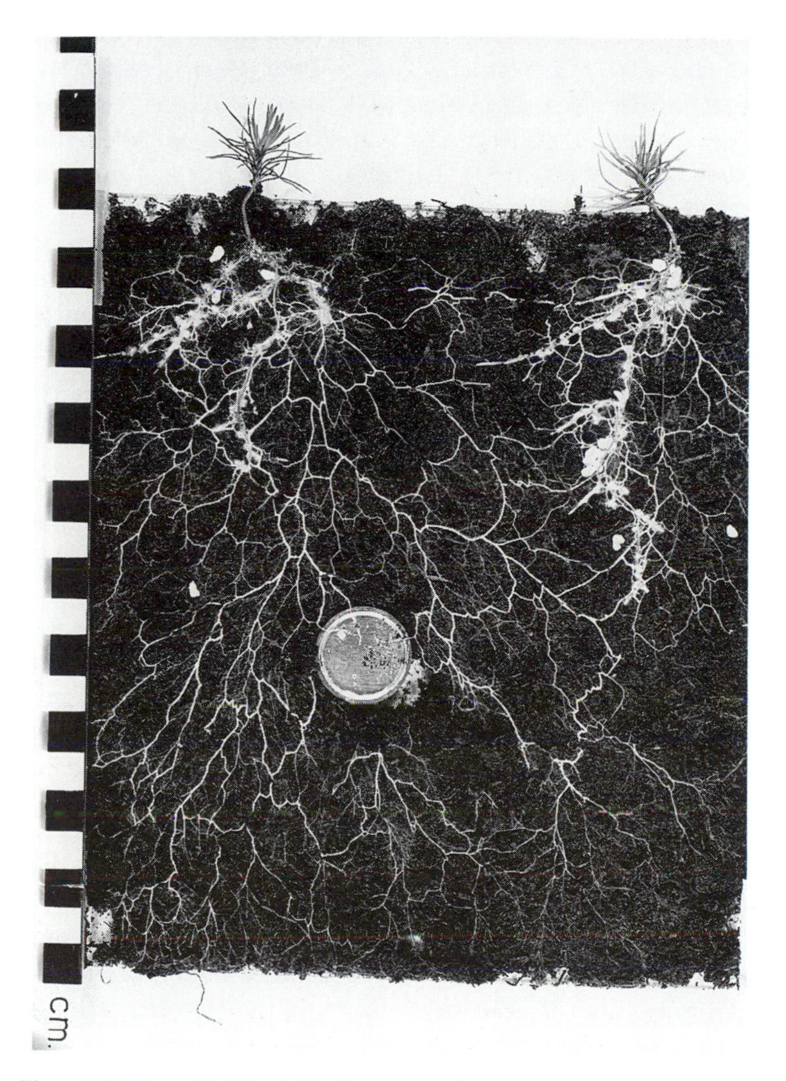

Figure 4.7. Mycorrhizal hyphal bridges connecting plants and the major routes of carbon and nutrient flow. As is shown, the vast majority of carbon from at least most connected plants flows to the fungus maximizing its own energy gain as opposed to connecting only one plant (photograph provided by David Read).

of VA mycorrhizal fungi contained ^{14}C from their host plants, and more recent studies have demonstrated that the bulk of the fungal carbon was probably derived from the host although limited use of external carbon sources was evident (see Cooper, 1984; Harris & Paul, 1987, for reviews).

Some extremely limited saprophytic growth of VA mycorrhizal fungi has been observed (e.g. Warner & Mosse, 1980; Hepper, 1983). Hyphae and spores of VA mycorrhizal fungi have been observed in the rhizosphere of nonmycotrophic plant groups (e.g. E. Allen & M. Allen, 1980, 1986; Schmidt & Reeves, 1984). Utilization of root exudates or limited and ineffective colonization has been hypothesized (Miller *et al.*, 1983). However, M. Allen (1983) demonstrated that arbuscules were frequent in perennial *Atriplex gardneri* in early spring during rapid leaf expansion, and M. Allen *et al.* (1989a) have shown arbuscules in nonmycotrophic annual seedlings. Using natural carbon isotope ratios, they also demonstrated that the VA mycorrhizal fungi were deriving most of their carbon from plants in these 'nonmycotrophic' plant groups (M. Allen & E. Allen, 1990).

Thus, despite the potential for limited carbon gain by mycorrhizal fungi from sources other than the host, at least in the field, most of these fungi approach obligate dependence on host plants to satisfy their energy and nutrient requirements.

Many mycorrhizal fungi associate with numerous plant species and interconnect nearby plants. The connections are usually regarded in the context of interplant dynamics. The resultant plant dynamics will be discussed in later chapters. However, the major function of these hyphal bridges may have little to do with the plants. The fungi are simply maximizing their own carbon gain (Figure 4.7).

Read *et al.* (1985) and Newman (1988) have outlined the evidence for interplant connections. The interpretation of these results for plant fitness is still controversial (see Chapter 5). However, the fungus always gains. In a mixed C_3–C_4 grassland, M. Allen *et al.* (1984a) found that a single fungal species, *Glomus fasciculatum* was associated with both *Agropyron smithii*, a C_3 grass, and *Bouteloua gracilis*, C_4 grass. However, the two grasses divided their phenology in such a manner that competition was reduced; the C_3 grass grew and photosynthesized early in the growing season whereas the C_4 grass grew and photosynthesized later into the hotter, drier part of the summer. The fungal activity was tightly coupled to the phenology of the host, active (in terms of having appropriate transport structures) at the same time as the plant. Thus, this fungus could gain carbon from both plants throughout the growing season depending on when each was active. M. Allen & E. Allen (1990) also recently looked at the $\delta^{13}C$ ratios of spores collected from the rhizospheres of mixed stands of C_3 grasses and shrubs, and C_4 chenopods. They found that the spores contained intermediate ratios, indicating that these fungi

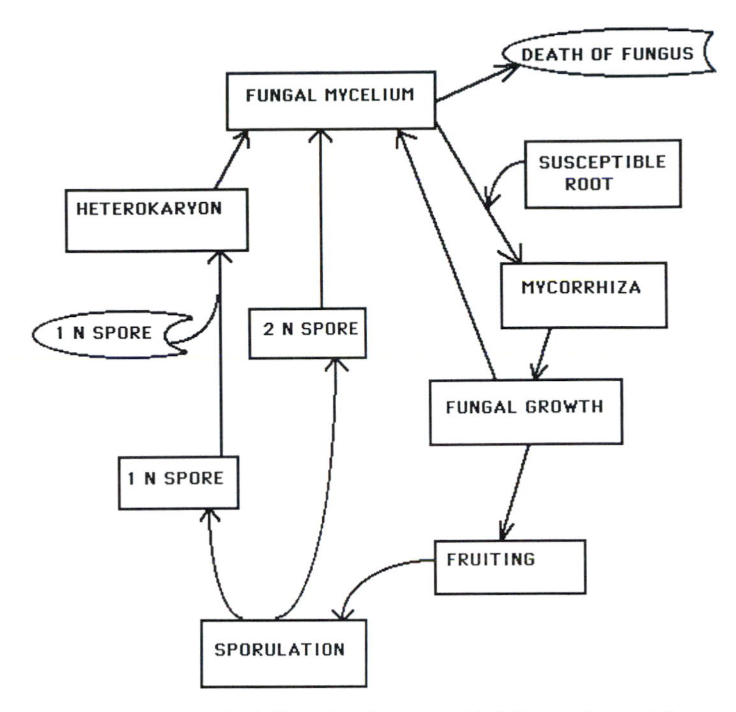

Figure 4.8. A generalized life cycle of a mycorrhizal fungus (see text for details).

were using carbon from both plant groups. At this time, it is not possible to measure the $\delta^{13}C$ ratio of an individual spore. Such information would determine whether or not a fungal 'individual' receives carbon from more than one plant.

Fitness of the fungal symbiont

Few fungi, including those that form mycorrhizae, have discrete generations or discrete individuals upon which most population modeling is based (e.g. May, 1981). The generalized life cycle of these fungi is not complicated in most cases, but responds in a highly plastic manner to its surrounding environment (Figure 4.8). An individual spore (sexual or asexual), whether surviving a stress period or dispersing to a new habitat, will germinate and the hyphae will expand as long as conditions for growth are adequate. This mycelium is analogous to an individual in plants. This 'individual' can be either haploid or diploid with each cell containing either one or more nuclei. If it is a VA mycorrhizal fungus or ectomycorrhizal *Endogone*, a Zygomycete, there are no regular cross-walls and therefore each hypha will contain numerous nuclei scattered along the hypha (Figure 4.9). The hyphal network, or mycelium, continues to expand until it reaches an unfavorable habitat or contacts another, compatible 'individual'. Following this contact, anastomosis may occur followed by exchange of genetic material. The genetic material may either remain intact as separate nuclei (if the nuclei are the same, it is referred to as a homokaryon; different, a heterokaryon) or fuse to become diploid. Rarely, the fungus continues to grow with a 2N nucleus. More often, it undergoes sexual recombination to produce sexual spores, or parasexual recombination to form new, recombined haploid nuclei (see any general mycology text for details). All three possibilities occur in mycorrhizal fungi, depending on the classification and genetic possibilities of each species.

Fitness in fungi is most readily defined in terms of survival through time of a population without enumerating offspring. This survival can take one of three forms: the ability to retain a patch of soil through time, the ability to tolerate stress conditions and re-establish following that stress period, and the ability to migrate and establish new colonies in new habitats.

Mycorrhizal fungi can retain a patch of soil for long time periods. Some basidiomycetous mycorrhizal fungi form fairy rings that can last up to hundreds of years. Saprophytic fungi have been shown to have substantial inertia once established in a site. For example, many tropical fungi can still be found in temperate grassland soils deposited during the altithermal and remnant tundra can be found in buried soils deposited during the Pleistocene (Christensen, 1989). The survival of the VA mycorrhizal fungus *Acaulospora* in cool temperate grasslands may be a similar phenomenon.

The survival of mycorrhizal fungi during perturbation has been better documented. Along the Pacific Coast of North America, some mycorrhizal fungi form arbutoid mycorrhizae with early successional manzanita and ectomycorrhizae with the later successional conifers (Perry *et al.*, 1989). Since the manzanitas resprout following a fire, these fungi would have a greater fitness than the specialists when attempting to survive and reestablish following a fire. Allen & Boosalis (1983) found that *Glomus fasciculatum* could survive annual losses of photosynthetic input. However, when the host changed from native perennial grasses to wheat that was planted only every other year with bare fallow in between plantings, the small-spored *G. fasciculatum* disappeared and was replaced by the larger-spored *G. mosseae* and *macrocarpum*. The *G. fasciculatum* presumably could survive annual stress periods but was unable to survive

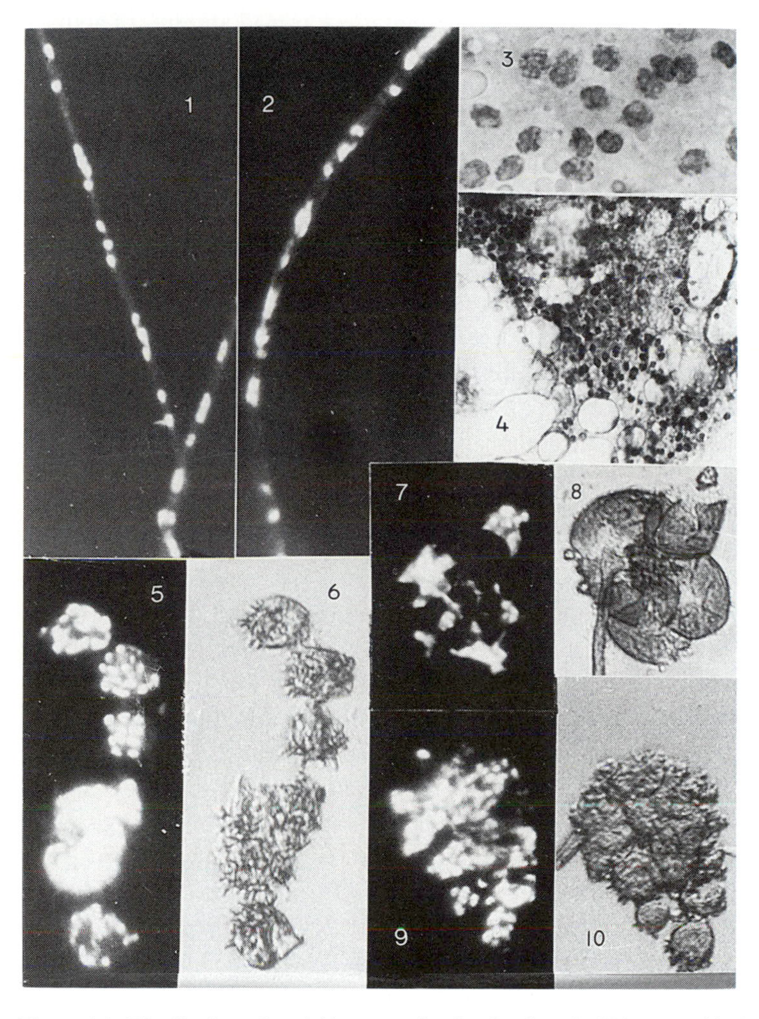

Figure 4.9. Distribution of nuclei in a germinating hypha of a VA mycorrhizal fungus as indicated by DAP1 fluorescence of the nuclei (from Cooke *et al.*, 1987, with permission). Notice the lack of cross walls and general migration of the nuclei down the hypha.

for 2 years without a host when competing with invading VA mycorrhizal fungi. Alternatively, in theory, spores of many fungi have the capacity to survive indefinitely. In areas of the Mount St Helens volcano where the original soil could be exposed via animals or erosion, both VA and ectomycorrhizal fungi survived and were abundant (M. Allen *et al.* 1984b; Carpenter *et al.*, 1987). In a Wyoming uranium mine, topsoil material stored for up to 12 years still contained a small amount of active inoculum (Christensen & Allen, 1980). The ability of these structures to survive and initiate new mycorrhizal infections is extremely important in the long-term survival of the symbiosis.

Dispersal and establishment of mycorrhizal fungi into new habitats

In order for mycorrhizal fungi to expand their range, there must be a source of new habitats. Disturbances may represent the major source of these new habitats available to mycorrhizal fungi. Many of the small disturbances such as those created by animals still retain some existing inoculum, allowing those fungi a better chance at establishment based solely on inoculum density. Severe disturbances, be they natural such as landslides or fire, or man-induced, such as agriculture or mineral extraction, create open habitats capable of invasion by new plants and mycorrhizal fungi. Although several fungal organs are capable of dispersing, most studies have centered on spore migration or organism distribution to understand dispersal mechanisms of mycorrhizal fungi.

Because these fungi are primarily dependent upon a host plant for energy, they must find a host for establishment and survival. As individual plants do not survive forever, and roots continually search out unoccupied soil, the ability of mycorrhizal fungi to find a host is limited.

Read *et al.* (1976) suggested that most VA mycorrhizal fungi spread via root-to-root contact. As individual roots grow through the soil, the hyphae follow these until another root is contacted, whereupon new infections are initiated. Powell (1979) estimated that VA mycorrhizal fungi could spread at a rate of up to 30 cm per year by root expansion in the glasshouse, ignoring seasonality. Warner & Mosse (1980) found that their VA mycorrhizal fungi expanded a few dm per year into sterile plots in the field. This rate of expansion could initiate new infections in small disturbances such as animal diggings. Reduced inoculum in gopher mounds has been observed in serpentine grasslands (Koide & Mooney, 1987) and forest clearings (M. Allen *et al.*, 1984b). Since hyphal expansion is too slow for colonization of larger-scale disturbances, spore dispersal must be examined.

Several ectomycorrhizal fungi fruit above ground; many of the forest mushrooms observed are mycorrhizal fungi. These fungi can disperse long distances if their spores can be entrained in the upper air flows above the canopy. However, turbulent bursts capable of reaching the forest floor to initiate dispersal may not be frequent enough to disperse spores

Figure 4.10. A squirrel feeding on an ectomycorrhizal sporocarp in a northern New Mexico forest. In the forest with limited wind penetration, turbulence is limited at the forest floor and spore dispersal via wind is limited. However, frequently, squirrels are observed feeding on the brightly colored sporocarps. This increases the entrainment potential as the turbulence is substantially increased (the Reynolds number increases from ~ 100 at the forest floor to 4×10^5 in the trees where squirrels feed).

dependably from fungi that fruit for only short periods. Allen (1987a) noted that the Reynolds number at forest floors was low, indicating limited turbulence-induced dispersal. Turbulence at the edge of forest clearings was much higher and spores of *Thelephora* sp., fungi observed growing near the forest edges, were found across the Mount St Helens eruption site. Many forest floor mycorrhizal fungi produce brightly colored fruiting bodies upon which squirrels and other small mammals frequently feed. When the squirrels feed, they often carry the fruiting structure into the trees where the turbulence is adequate for entrainment into the upper air flows (Figure 4.10). This might act as a selective mechanism for the bright coloring of these mushrooms, assuming that small mammals are attracted by colors.

Many other ectomycorrhizal fungi and all VA mycorrhizal fungi fruit below ground (commonly referred to as hypogeous fungi). These fungi have been observed to spread via root expansion, via animals or via some erosion agent. Several workers have observed interplant connections of both VA and ectomycorrhizal fungi (see Read *et al.*, 1985). The detection of truffles by pigs and dogs is well known. These fungi release pheromones attractive to many mammals (including humans). This could increase their long-range dispersal potential (from Italy to California, if the price is right). Maser *et al.* (1978a) have reported that up to 80% of the diets of rodents were composed of hypogenous fungi (predominantly mycorrhizal), and Allen & MacMahon (1988) have experimentally determined that this rodent inoculum could result in the spread of mycorrhizae across the Mount St Helens volcanic eruption landscape.

Erosion represents a major dispersal pathway for mycorrhizal fungi and one in need of more study. Trappe (1988) and Watling (1988) have observed mycorrhizal fungal fruiting bodies downstream from vegetative communities of origin. Erosion zones in the tephra of volcanic eruptions are a source of mycorrhizal inoculum (M. Allen *et al.*, 1984b; Carpenter *et al.*, 1987). Friese (1984) observed that *Ammophila* plants that were eroded from the beach, washed out to sea, and redeposited on the high tide lines, contained VA mycorrhizal inoculum. In arid habitats, wind erosion moves VA mycorrhizal spores up to 2 km (Warner *et al.*, 1987).

Fitness and establishing new mycorrhizal associations

For mycorrhizal fungi to establish, either from localized contact or from dispersed inoculum, the fungus must invade and colonize a host root. Four factors affect the formation of mycorrhizal infections: host and fungal genetic compatibility, edaphic factors, plant–microbial activity, and inoculum density. These factors comprise what Garrett (1970) defines as 'inoculum potential'. These factors can be broken down into various components, but if the fungus does not establish, it simply will not survive or reproduce.

Mycorrhizal fungi have a wide range of specificity responses. Some, such as many species of *Glomus*, show almost no specificity within plant species capable of forming VA mycorrhizae. In fact, although there are considerable differences in host responses to differing VA mycorrhizal fungi, no reports are available of invasion specificity of VA mycorrhizal fungi. Curtis (1939) noted that, in his samples, little specificity was apparent among orchid mycorrhizal fungi. Alternatively, Warcup (1981) found that specificity was highly variable; some Australian associations (e.g. *Caladenia* and *Sebacina vermifera*) are highly specific, whereas others may be specific at the species to subtribe level. Molina & Trappe (1982a) reviewed the wide range in specificity of ectomycorrhizal fungi from the highly specific *Alpova diplophloeus* with *Alnus* to the numerous widely distributed fungi with wide host-conifer range (e.g. *Amanita muscaria* and *Boletus edulis*). One broad host fungus, *Pisolithus tinctorius*, has been spread world-wide for use in forestation (Marx *et al.*, 1984).

The specificity of mycorrhizal fungi is critical to any discussion of the fitness of that fungus. Fungi that are highly specific often come from specialized habitats. Alder rapidly invades disturbed subalpine wet habitats and remains as a dominant plant without replacement until a new, major perturbation (Trappe, 1988). Its mycorrhizal fungi tend to be highly specific (Molina & Trappe, 1982a), with the result that these fungi can survive as long as the alders remain. Many ericoid mycorrhizal fungi associate with a wide diversity of heathland shrubs that tend to be highly stable through time, but do not form mycorrhizae with other plants. Forests such as those in the Pacific Northwest undergo transitions between dominant plants. Interestingly, much of that shift occurs between ericaceous shrubs such as *Arctostaphylos uva-ursi* and trees such as *Pseudotsuga menziesii*. Several mycorrhizal fungi (e.g. *Rhizopogon vinicolor* and *Suillus lakei*) form arbutoid mycorrhizae with the ericaceous shrubs and ectomycorrhizae with the coniferous trees (Molina & Trappe, 1982b) (Figure 4.11). This insures these fungi a host despite changes in the plant community, as well as access to multiple carbon sources (discussed later). The fungi that form VA mycorrhizae show little specificity. VA mycorrhizae form on herbs as well as shrubs and trees. The ability to find a host upon the death of a single herbaceous individual could be critical to the survival of these fungi; thus specificity may have been selected against.

Edaphic factors affect the ability of any organism to survive. Because a mycorrhiza is a composite of two organisms, any edaphic factor that affects one affects the other. High available nutrient levels and saturated soils often inhibit mycorrhizal symbioses (see Chapter 1). One general assumption is that if the plant has access to all its necessary resources, the symbiosis will not form as the fungus then becomes a carbon drain (see earlier discussion). Another alternative is that the fungus will invade any root encountered (see chapter 3). Both hypotheses assume that the plant is the regulating agent. If the plant has adequate resources or if the fungus is not a compatible type, then the plant will reject the symbiosis, but if it is deficient it will accept the symbiosis. The fungus continues to invade as long as it has adequate energy. The details of this interaction, however, remain controversial and hazy. This would be an excellent research topic.

Since the role of edaphic factors is not well understood, it is difficult, at best, to delineate the roles of other organisms in affecting mycorrhizal

Figure 4.11. A successional stand of manzanita with arbutoid mycorrhizae and Douglas fir seedlings with ectomycorrhizae. Many fungi forming these mycorrhizae are the same species. Since the manzanita resprout following fire, these plants may retain a reserve of mycorrhizal fungi that aid in the reforming of ectomycorrhizae that enhance conifer establishment (Amaranthus & Perry, 1989).

infection. The potential for site competition between root-penetrating organisms and mycorrhizal fungi needs further study. Kowalski (1977, 1980, 1982), Hetrick (e.g. 1984), Ames (1987) and others have provided some evidence that interference from other root-inhabiting micro-organisms may affect mycorrhizal formation, but additional evidence for differing systems is needed. The activity of neighboring plants is known to affect mycorrhizal formation. Competition between plants, in part a function of plant density, determines the resources available. If a high plant density exists, resources will be lower and, just as low soil nutrients concentrations can stimulate infection, increasing competition can result in resource depletion and increase infection (E. Allen & M. Allen, 1980). Finally, some evidence suggests that the plant itself affects the infection process. Fries & Birraux (1980) found that living plant roots stimulated germination in *Hebeloma*. Koske (1982) and Gemma & Koske (1988) showed that volatiles from a host plant attracted the hyphae of *Gigaspora*. Further work on the mechanisms of these actions and the breadth of these responses is needed.

General summary

Mycorrhizal fungi require carbon for a host plant to survive. They have numerous strategies for obtaining carbon ranging from the specialists living on limited plant species in particular habitats with long survival times, to generalists found in swards or associated with early- and later-successional plant groups. Mycorrhizal fungi, as is the case for most fungi, have a high diversity of genetic characteristics and reproductive structures that change and exchange in complex ways, making it difficult to quantify fitness. However, they have an extraordinary capacity for growing, dispersing, and surviving stress periods. These abilities make them highly successful organisms despite their dependence on another organism for growth and reproduction.

Several factors operating in concert have created a symbiosis that is both widespread and has lasted through much of the history of terrestrial life. Why can such disparate organisms successfully live together when theoretical models suggest that such symbioses should not be successful? In part, this success may arise because few mycorrhizae represent highly specific relationships. Most plants have a wide variety of fungal partners and most fungi have multiple hosts. Thus, if one partner dies, there are many others from which to draw resources. Mycorrhizal associations also have the capacity to expand the resource acquisition rate; in addition, the association provides access to resources unavailable to the individual organism.

Both plants and fungi have complex means, sexual and asexual, of surviving and spreading their genes. They also both have 'individuals' that spread, undergo genetic recombination, and die, although these 'individuals' are often difficult to detect. Thus, theoretical models designed to describe the characteristics and potentials for mutualisms should relate to these organism as well as to 'higher' animals if they are truly global.

Finally, the mycorrhiza is more than a single factor mutualism that evolved only to take up phosphorus. A mycorrhiza, composed of two organisms that provide needed resources to one another, affects the fitness of each of the partners. Large responses may not always be measured, but an increase provided by the fungus in the survival of a plant population through a single stress period may be enough to improve the fitness of that plant. It is in this context of contributions to fitness, not simply production, that we must strive to understand the evolution and ecology of mycorrhizae.

5

Community ecology

Mycorrhizae affect all organisms that utilize the same resources as either symbiont or that use either symbiont as a resource. An individual plant acquires resources from mycorrhizal fungi, enabling that plant to grow faster, produce more offspring, and be more competitive. The mycorrhizal fungus has a direct pipeline to high energy carbohydrates of a healthy plant, a distinct advantage compared with a saprophytic fungus that must rely on dead organic matter consisting of complex and often highly toxic molecules or a parasite which must foil a host's defenses. Mycorrhizal fungi and mycorrhizal plants serve as food for a variety of animals, ranging from nematodes to humans. Fungus and plant also host numerous parasites of all sizes. By altering the resource base of some organisms and becoming a resource for others, the mycorrhizal symbiosis contributes to the structure and functioning of the surrounding community. Because they are so abundant in most communities, it can be argued that they regulate the functioning of those communities.

The role of mycorrhizae in community ecology is difficult to describe because delineating a single 'microbial community' is virtually impossible. In defining a plant or animal community, human perception constructs borders, and the land unit inside those borders becomes a community. For example, many plant community types are well recognized, named entities, especially in European vegetation ecology (e.g. Falinski, 1986). Individual plant communities may also be distinguished statistically wherein the variation in species compositions between relevés are greater than some accepted level (Mueller-Dombois & Ellenberg, 1974). Saprophytic fungal communities are generally described by their relationships to some plant community or are referred to as microcommunities living on one particular substrate. Saprophytic fungi are similar in species composition between plant communities containing similar dominant plant species even if they are separated by continents (e.g. Christensen,

1981). Nevertheless, they may also be as distinct in two different microhabitats near the same plant as between biomes. For example, Allen & MacMahon (1985) found that in a cold desert shrubland, the dominant fungi in the interspace of a shrub were desert fungi whereas those under the shrub were tundra–taiga species. The species overlap value (using Sorenson's index) was only 0.22 between fungi under a single shrub and its associated interspace.

Spatially, mycorrhizal fungal communities are not yet understood. One fungus can link several host plants (e.g. Warner & Mosse, 1980; Read, 1984), but because the hyphae are so fine, an individual mycelium will never occupy all of the available space within that matrix. Statistical treatments have described the spatial structure of VA mycorrhizal fungal spores as being aggregated at virtually all scales tested (Anderson *et al.*, 1983, using 625 cm^2 plots; Friese, 1984, using 1000 cm^3 plots; Sylvia, 1986, using cores taken at 33 cm intervals across a transect; Allen & MacMahon, 1985, at 2 cm distant cores in a grid; St John & Koske 1988, random cores). VA mycorrhizal root infections (St John & Hunt, 1983) and hyphae (Allen & MacMahon, 1985) are also aggregated and often correlated with soil characteristics such as organic matter and nutrients. Ectomycorrhizal fungi have not been described using these approaches but these fungi can link relatively large land units that encompass several plants (e.g. Ogawa, 1985). The same fungus that comprises an ectomycorrhiza can also link and form an arbutoid mycorrhiza with a neighboring plant (Perry *et al.*, 1989). This might suggest that the mycorrhizal fungal community can be defined at a scale that is independent of either the plant as an individual or the plant community as a whole.

One approach describes the community dynamics based on the organism; that is, it defines community processes in the context of a mycorrhizal association rather than delineating a 'mycorrhizal community'. To take this approach, the definition of a community proposed by MacMahon and colleagues (MacMahon *et al.*, 1981) will best serve as a working model; a community is:

> the organisms which affect, directly or indirectly, the expected reproductive success of a reference organism.

Using this approach, I will review the literature that deals with responses of the community to mycorrhizae including literature that defines resource capture mechanisms, competition for resources both among and between plants and fungi, and other parameters that control community functioning. The difficulty becomes scaling up from interactions between two organisms to describe the dynamics of an entire community. Nevertheless, if the processes can be understood, then the patterns become interpretable.

Responses of plants to mycorrhizae have been studied almost exclusively in glasshouses in pot cultures consisting of one fungus and one plant. Therefore, although a major interest in mycorrhizal research is how the symbiosis increases resource acquisition, few data exist on how mycorrhizae affect community composition or functioning. Because plant production in the field does not occur in the absence of neighbors, understanding the roles of mycorrhizae in communities is essential to understanding how the symbiosis affects plant dynamics.

Fungal community interactions

Mycorrhizal fungi can regulate the communities in which they occur by three types of interactions. Each is scale- and time-dependent. These include (1) direct contact between a hypha of the mycorrhizal fungus and the other organism, (2) modifications of the environment by one organism, e.g. contact between one organism and the products of another, and (3) indirect influence via a third organism, in this case the host plant.

The first, or direct contact between microorganisms, is easiest to understand. Microorganisms in search of resources, such as mycorrhizal fungi, come into contact with competitors, predators and parasites as do other organisms. Mycorrhizal fungi compete for space along a root and soil nutrients with other mycorrhizal fungi, as well as with parasitic and saprophytic fungi, bacteria and plant roots. A fungus must also compete with itself. As a hypha expands into the substrate, it branches and the resulting hyphae take up resources, creating depletion zones around each mycelial system. VA mycorrhizal fungi, and probably others, form sets of absorbing hyphae branching out from a single runner hypha. These are presumably formed at given intervals in front of depletion zones formed by the prior absorbing hyphal matrix (Figure 5.1). Unfortunately, most of these interactions occur at scales difficult to measure. Nevertheless, they occur frequently in a soil matrix, making these interactions critical to plant resource acquisition.

Direct predation and parasitism are well known for mycorrhizal fungi. A wide variety of invertebrates feed on the hyphae of mycorrhizal fungi, and animals, including many large vertebrates, eat mycorrhizal sporocarps

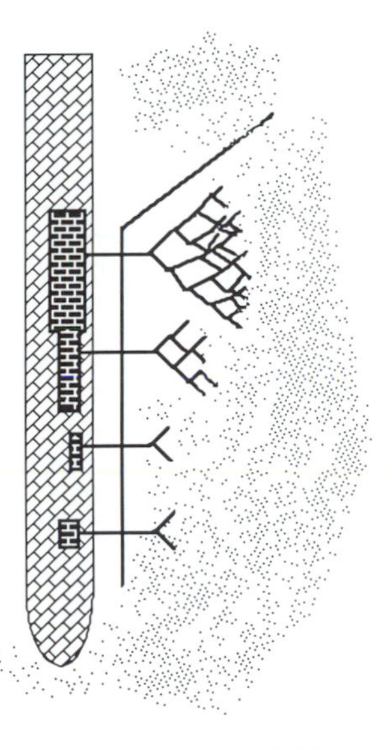

Figure 5.1. Structure of VA mycorrhizae and the resulting depletion of soil resources. These patterns are especially important for resources that are relatively immobile in soils such as phosphorus and ammonium but in severely limiting conditions could apply to almost any limiting nutrient. Shown are the root (▨), the root occupied by the mycorrhiza (▤), the mycorrhizal hypha (—<) and the limiting nutrient (░).

(see animal–fungal interactions, later in this chapter). Bacterial and fungal parasites have been observed, especially upon storage of mycorrhizal fungal propagules (e.g. Daniels-Hetrick, 1984; Paulitz & Menge, 1986), and are thought to be important in the distribution and effects of mycorrhizae in the field (e.g. Ross & Runnencutter, 1977; Koske, 1981). Mycorrhizal fungi also act as animal predators; *Glomus fasciculatum* has been reported as a pathogen of the nematode, *Heterodera glycines* (Francl & Dropkin, 1985).

The second interaction, effects of substances or depletion zones created by one organism on a competitor, is widespread and is determined by the nutritional modes of fungi and other microorganisms. Fungi, including mycorrhizal fungi, gain their nutrients by releasing enzymes into a substrate. Those enzymes break down complex molecules into simple resources that can be absorbed directly. This also creates a free energy gradient, a high concentration where the complex molecule was present, and a low concentration along the hypha where the nutrients were absorbed and translocated. The nutrients presumably diffuse continuously from the point of action of the enzyme to the fungus depending on

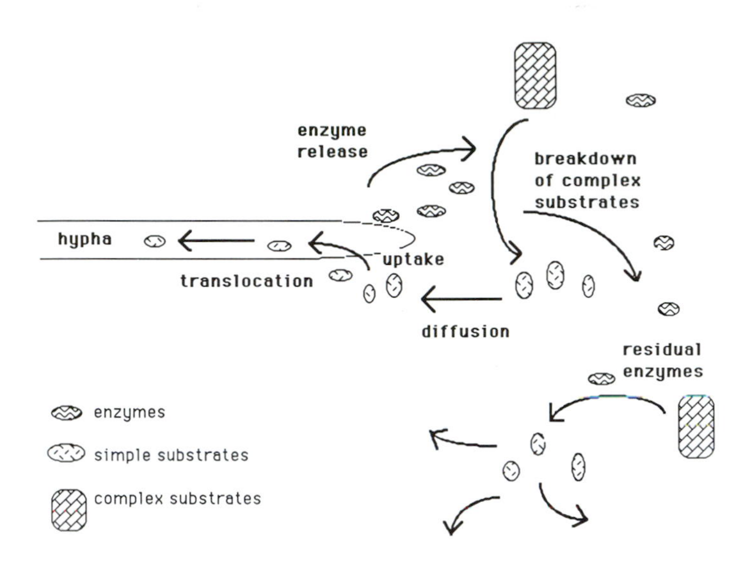

Figure 5.2. Mechanisms of nutrient gain by fungi (absorptive nutrition) and the means by which fungi (including mycorrhizal fungi) can affect nearby organisms without coming into direct contact. Enzymes are released into the substrate that alter the chemical composition of the environment, often remaining after the organism is gone.

substrate conditions, the absorbing capacity of the organism, and the diffusion capacity of the ion species (Figure 5.2). These enzymes produce depletion zones beyond the organism making microsites undesirable for habitation by other organisms.

Mycorrhizal fungi produce and are sensitive to antibiotics, allelopathic compounds produced by microorganisms to gain and control resources. These compounds are released into the environment, affecting organisms away from the releasing fungus. An example is chloromycorrhizin A from a monotropoid mycorrhizal fungus (Stalhandski *et al.*, 1977). The potential for persistence and accumulation of these compounds from soil fungi is well known, possibly accounting for the 'fungistasis' of some soils.

The third interaction, via the host plant, is potentially of the greatest importance to these fungi. Mycorrhizal fungi influence each other as well as other fungi by occupying discrete regions of a root or by occupying specific root cells. Zones of mycorrhizal fungi rarely host other microorganisms although the mechanisms of interaction are not known. Possibly a generalized triggering of a host's defense (discussed in Chapter 4) needs to be further addressed. Mycorrhizal fungi improve their own survival by increasing the fitness and survival of their host to the detriment of hosts of other mycorrhizal fungi. This mode of interaction should be especially important in highly specific associations.

Despite efforts to establish a 'definitive role of mycorrhizae' in communities, the fungi comprising mycorrhizal symbioses are diverse taxonomically, morphologically, and physiologically. For example, Trappe (1977) estimated that more than 2000 species of mycorrhizal fungi are associated with Douglas-fir alone! Ford *et al.* (1980) observed changing patterns of sporophores of three species of ectomycorrhizal fungi around a single young birch tree over a 3-year period. In one 100 m transect, five genera and as many as 15 species of VA mycorrhizal fungi, many undescribed, were found in a sagebrush (*Artemisia tridentata*) community with only eight plant species (C. Friese and M. Allen, unpublished data). Sylvia (1986) found three genera and seven species of VA mycorrhizal fungi in a single plant species, *Uniola paniculata* (sea oats), on a pioneer dune field in Florida and Friese (1984) found eight viable species of VA mycorrhizal fungi associated with a single individual of *Ammophila breviligulata* in a Rhode Island dune.

Mycorrhizal fungi show a high degree of physiological diversity even among fungi that live on the same host or fungi that exhibit little host specificity. Mexal & Reid (1973) found wide variation in drought tolerance of some generalist ectomycorrhizal fungi. Allen & Boosalis (1983) found that the two co-occurring *Glomus* species affected the water relations of wheat (*Triticum aestivum*) differently, in that one species (*G. fasciculatum*) improved the drought tolerance of the host whereas the second species (*G. mosseae*) reduced its drought tolerance. Gildon & Tinker (1981) isolated strains of *Glomus mosseae* with different capacities for tolerating heavy metals. Stahl & Smith (1984) reported that both different ecotypes and species of *Glomus microcarpum* and *macrocarpum* differentially alter water relations of *Agropyron smithii* regardless of whether the fungi were isolated from the same or different sites associated with the same host. Graham *et al.* (1982) and Abbott & Robson (1985) found that different fungal species formed different hyphal lengths per infected root length that may be related to the different nutritional effects of these species on the host plants.

These data together indicate that there is a high diversity of mycorrhizal fungi often associated with the same plant. Thus, there can be no 'single mycorrhizal effect' on plant communities.

Most surveys of mycorrhizal fungi have occurred across communities and have demonstrated both differences among fungal species and mycorrhizal types. Different habitats containing different plant species are well known to have different mycorrhizal associates (e.g. Read *et al* ., 1976; Singer *et al.*, 1979; Read & Haselwandter, 1981; Miller, 1987; Janos, 1987). Individual ectomycorrhizal fungi change with hosts and edaphic conditions. For example, *Tuber* spp. varied with hosts and across soils and moisture gradients (Zambonelli & Govi, 1983; Zambonelli & Morara, 1984). Some VA mycorrhizal fungi, such as *Gigaspora margarita* and *Glomus fasciculatum* may have the widest distribution of any species worldwide although there is considerable controversy as to the taxonomic status of many of these species (e.g. Walker, 1985). Fungal species shifts are also common with broadly ranging plant species. In Wyoming sagebrush grasslands (*Artemisia–Agropyron*) species of *Glomus* dropped out and the total spore numbers decreased with increasing aridity (Stahl & Christensen, 1983). Pond *et al.* (1984) found several monospecific stands of VA mycorrhizal fungi in different habitats but no correlations were apparent with salinity or host species. Koske and colleagues (e.g. Koske, 1981, 1987; Koske & Halvorson, 1981; Bergen & Koske, 1984) have provided an extensive data base demonstrating changes in VA mycorrhizal fungal species associated with *Ammophila brevigulata* over gradients along several hundreds of miles. Interestingly, few correlations appear to be found with nutrients, organic matter or other variables that are normally suggested to be the primary determinants of fungal distributions. Koske (1987) suggested that the shifts in species compositions appear to be primarily climatological.

However, mycorrhizal fungi also segregate out spatially within at least three scales in a community. These scales occur with respect to environmental heterogeneity, host specificity, and architecture of a root system. Within a community, Ebbers *et al.* (1987) reported that soil moisture–nutrient gradients were associated with fungal species distributions across a prairie. In a detailed examination of fungal species associated with a single plant species (*Sporobolus heterolepis*), the plant and fungal species did not necessarily coexist along the same environmental gradients, suggesting that they were responding to different variables. A high degree of host specificity appears to occur in specialized habitats where, following the establishment of a given plant species, it tends to persist. An example of this pattern is in *Alnus* spp. growing in bottom land habitats (Molina, 1981; Molina & Trappe, 1982a).

In many patches within a plant community, mycorrhizae tend to differentiate according to successional state or slight habitat differences. In sites where succession results in species replacement, many of the fungi

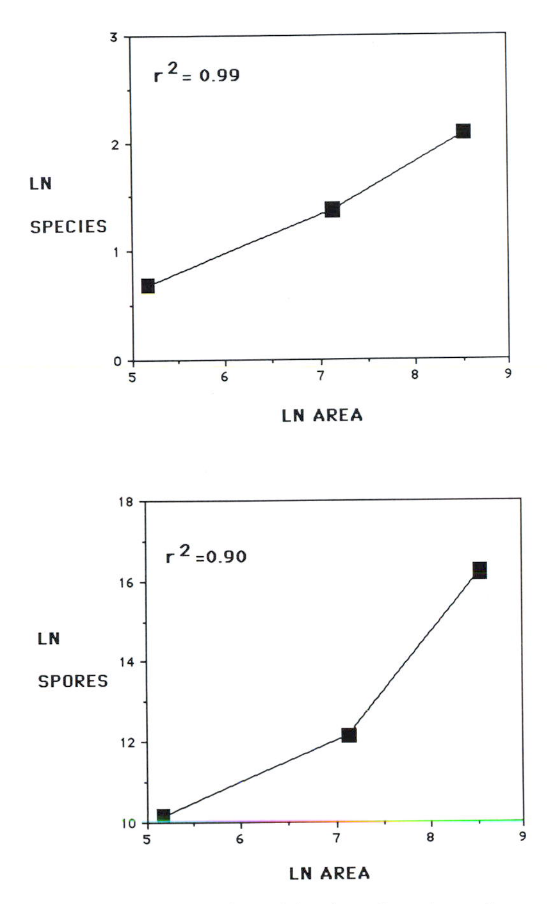

Figure 5.3. Relationship of density of species and spores of VA mycorrhizal fungi to area (horizontal root extension) of a host where the individual plants are far enough apart so that the root systems are not in contact with neighboring individuals. These data suggest that the area of the individual and/or the carbon input regulates the activity of the mycorrhizal fungi in a way that habitat size or diversity regulates the populations of other organisms (from M. Allen, 1988a).

are nonspecific. An interesting case is a fungus that forms arbutoid mycorrhizae with earlier seral Ericaceae and then ectomycorrhizae with establishing pines (Amaranthus & Perry, 1989). An individual fungus may associate with several plants even of different species (e.g. Molina & Trappe, 1982a; Chiarello *et al.*, 1982). Presumably, this would improve the carbon gain potential of the fungus. M. Allen & E. Allen (1990) have found that a spore population had carbon derived from both C_3 and C_4 plants although the carbon source for a single spore cannot yet be detected. The ability of the inoculum to grow with a wide range of hosts also is used to increase inoculum for a particular site. For example, Kormanik *et al.* (1980) increased VA mycorrhizal fungal inoculum for tulip tree (*Liriodendron*) by first inoculating with a small amount of VA mycorrhizal fungal inoculum and then expanding that inoculum with a cover crop such as sorghum.

In an arid habitat, where vegetation recovery was characterized by few seral stages or autosuccession, the mycorrhizal fungal species diversity appeared to reach equilibrium with the size of the individual plant associate (Figure 5.3). This suggests that the plant carbon input or the extent of root distribution limits the extent of mycorrhizal activity in the same way that island size affects the diversity of higher animals (M. Allen, 1988a). As a single plant expands outward, its roots encounter greater habitat diversity and the roots undergo structural changes that influence both their mycorrhizal and saprophytic microorganism associates (Figure 5.4). Thus, the architecture and size of an individual plant regulate the distribution and diversity of mycorrhizal fungi.

In most habitats, mycorrhizal activity decreases as the roots go deeper into the soil, often rapidly approaching no mycorrhizae (e.g. Sparling & Tinker, 1978; Buchholz & Motto, 1981; Schwab & Reeves, 1981; Tarasova & Dumikyan, 1984). However, notable exceptions do exist. Virginia *et al.* (1986) found mycorrhizal fungi down to 4 m in sandy arid soils. In this system, the roots penetrated to the buried water table and mycorrhizal activity was concentrated just above that region. More importantly, the species of mycorrhizal fungi may also segregate with depth. Zajicek *et al.* (1986) reported that in prairie soils, VA mycorrhizal fungal activity extended more than a meter into the soil but that the species richness declined with depth. Under *Artemisia tridentata*, a high diversity (4 species) of VA mycorrhizal fungi were found in the top 20 cm, but by 60 cm only *Glomus microcarpum* was left (M. Allen, unpublished data).

Horizontally, there appear to be major changes in mycorrhizal activity in mesic habitats, but little in the sparsely vegetated arid regions. For example, Ford *et al.* (1980) noted that the fairy rings of ectomycorrhizal fungi shifted positions around a single beech seedling through time but distinct zones of occupation were apparent. Fleming (1984) found distinct spatial partitions in the distribution of ectomycorrhizal fungi after growth in the field. Alternatively, Allen & MacMahon (1985) found that spores of a variety of fungi were continually concentrated around the base of a sagebrush plant in an arid site.

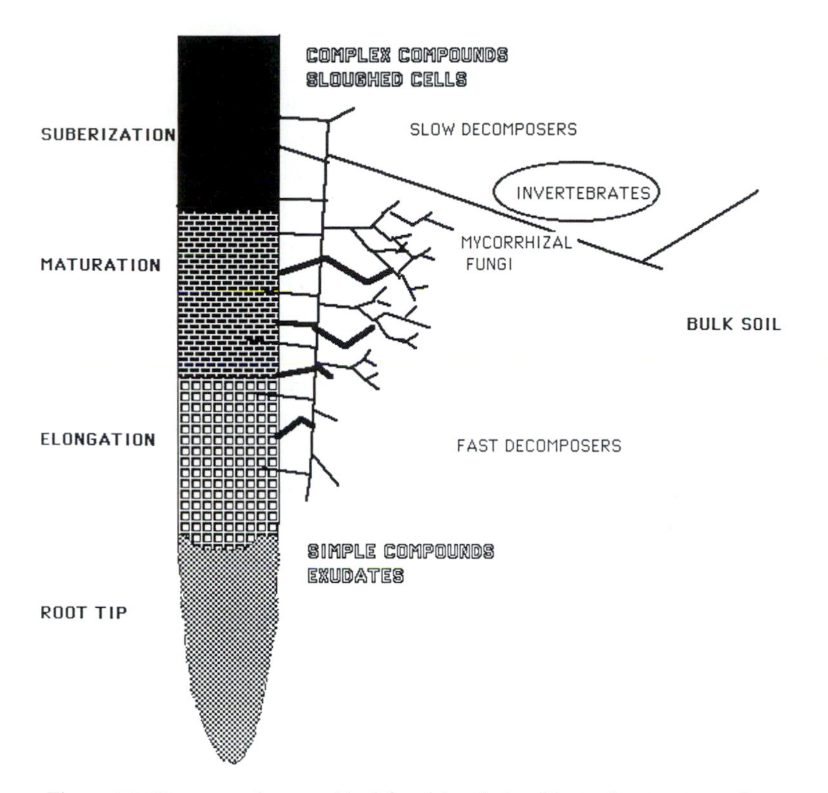

Figure 5.4. Structure of mycorrhizal fungi in relationship to the structure of the root and associated root organisms (derived from Coleman *et al.*, 1983).

Probably the most interesting interactions exist on single host plants that can form both endomycorrhizal and ectomycorrhizal associations. Plants in a wide variety of genera including such common cross-overs as *Populus*, *Salix*, *Casaurina*, and some ferns form both VA and ectomycorrhizae (see Trappe, 1987 for a comprehensive discussion). Although only a few authors have attempted to understand these interactions, such evidence as there is suggests that the initial colonizing fungi are VA mycorrhizal with the ectomycorrhizal fungi able successfully to invade those roots after VA mycorrhizae are active. The reverse (VA mycorrhizal fungi into ectomycorrhizal roots) appears not to occur (e.g. Chilvers *et al.*, 1987). E-strain ectendomycorrhizal fungi occur commonly on pines. These fungi appear often to be early colonizers of these roots and are replaced with ectomycorrhizal fungi. Again, the reverse direction, replacement of ectomycorrhizae by ectendomycorrhizae, does not appear to occur (e.g. Mikola, 1965).

Much replacement may be dependent on the site conditions. In the case of E-strain ectendomycorrhizae, these fungi are generally rapid colonizers in disturbed lands and are rapidly replaced by ectomycorrhizae in natural, high-fertility soils (Mikola, 1965). However, Mikola (1965) also noted that replacement was very slow on highly disturbed sites. LoBuglio & Wilcox (1988) found that the original inoculum type (E-strain or ectendomycorrhizae) often persisted without replacement in iron tailings. Similar patterns have also been observed in *Eucalyptus dumosa* seedlings as they grow, a kind of succession from endo- to ectomycorrhizal fungi (Lapeyrie & Chilvers, 1985).

The fungi forming any one type of mycorrhiza compete among each other for sites. Wilson (1984) suggested that the different competitive colonization of different VA mycorrhizal fungi could be accounted for by different inoculum potentials of the different fungi despite the ability of all species to associate with the host plants. Abbott & Robson (1984) reported that the initial colonizing VA mycorrhizal fungus had the definite competitive advantage. However, Hepper *et al.* (1988) found that some VA mycorrhizal fungi could invade a VA mycorrhizal root occupied by a different species but that the reverse was not necessarily true. Numerous examples exist of ectomycorrhizal fungal replacement during forest succession. The ability of *Thelephora terrestris* to colonize nursery-bed plants and its subsequent replacement by other ectomycorrhizal fungi following planting in the field is well known (Trappe, 1977). *Thelephora terrestris* rapidly establishes on plants invading disturbed habitats such as the Mount St Helens Pumice Plain, an area covered to 20 m with sterile ash following the 1980 eruption (M. Allen, 1987a). Other fungi, such as species of *Russula* are only found in older, established forests (e.g. Murakami, 1987; M. Allen, unpublished observations). Dighton *et al.* (1986) also documented the shifts in mycorrhizal fungi during succession by following the occurrence of fruiting bodies. Some evidence from the field suggests that many ectomycorrhizal fungi have r or K strategies. For example, Fleming (1985) noted that there were 'early-stage' and 'late-stage' fungi forming ectomycorrhizae in potted individuals of *Betula* spp. corresponding to r and K strategies because of their dispersal and competitive characteristics.

Fungal allelopathy as well as competition can affect the mycorrhizal fungal species composition which, in turn, can affect the higher plant community. Dimbleby (1953) suggested that the mycorrhizal fungi of heather might inhibit the development of ectomycorrhizae of pine and birch and thereby reduce the reforestation of moors. Brown & Mikola

Figure 5.5. Distribution of several species of ectomycorrhizal fungi within a single root system. The factors that govern this microscale distribution are not known but are presumably related to niche differentiation among the fungi and possibly competition for space.

(1974) hypothesized that fruticose soil lichens inhibited the ectomycorrhizae and consequently conifer seedling establishment. Tobiesson & Werner (1980) and Koviacic *et al.* (1984) noted that pine litter inhibited the growth and germination of VA mycorrhizal fungi and thereby inhibited hardwood and herb survival. Rose *et al.* (1983) documented the allelochemic effects of various tree litters on each other's ectomycorrhizal fungi with the low concentration for action being 10 ppm of allelochemical. Alternatively, Anderson & Liberta (1987) failed to find any evidence for allelopathic activity in early successional fields. As has been the case in the plant allelopathy literature, careful documentation of the required concentrations for action (frequently measured), mechanisms of action (sometimes measured), and active concentrations in the field (almost never measured) are essential to demonstrate these effects in mycorrhizal fungi.

As with other organisms, similar fungal species show niche differentiation. There is no such thing as a generic mycorrhizal fungus and mycorrhizal effect. In soils broken from prairie sod in the western Great Plains, *Glomus fasciculatum* was replaced by *Glomus mosseae* with negative effects on plant production (Allen & Boosalis, 1983). *Pisolithus tinctorius* is commonly used to reclaim severely disturbed habitats but it is generally replaced as the soil organic matter increases and other fungi invade (e.g. Marx, 1977). In nature this fungus is associated with *Pinus*, often in areas of low organic matter (e.g. coastal sand dunes: M. Allen, unpublished observations). The important roles of litter and organic matter on the distribution of ectomycorrhizal fungi have been documented experimentally (e.g. Ogawa *et al.*, 1981, 1983), and fertilization of a forest floor can change the mycorrhizal species population densities even if it does not change the compositions (e.g. Menge & Grand, 1981). Fungal associates can be seen to change as a single root grows through different substrates (Fleming, 1985). For example, roots from the same individual tree have different fungal associates in mineral soil as they do when growing into decaying logs (e.g. Harvey *et al.*, 1980). Most differentiation probably occurs at scales that have not yet been studied. If one observes an ectomycorrhizal root from a forest, several fungi can usually be distinguished along its length (Figure 5.5). Hypothetically, much of this pattern results from differing microscale soil conditions favoring the growth of one fungal species over another.

Mycorrhizal fungi interact with a wide variety of the soil and root fungi, especially plant pathogens. Some authors have suggested that the mycorrhizal fungi alter the host plant in such a manner as to affect these other fungi. For example, Davis & Menge (1981) found that improved P status provided by differing VA mycorrhizal fungal isolates reduced disease caused by *Phytophthora parasitica*. Graham & Menge (1982) suggested that VA mycorrhizal fungi increase plant P, which reduces root exudates, which in turn reduces root pathogens such as *Gaeumannomyces graminis*. Other mycorrhizal fungi, particularly those forming ectomycorrhizae, reduce pathogen activity by completely enclosing a root segment or by producing antibiotics (e.g. Marx, 1969a,b; Marx & Davey, 1969a,b; Schenck, 1981). In some of the most interesting studies, Kowalski (1977, 1980) showed that the soil saprophytic microfungi affected both ectomycorrhizae of *Pinus sylvestris* and the pathogen *Cylindrocarpon destructans* to the extent that disease development and plant survival were regulated in forest plantings in Poland.

Simultaneous attempts to gain soil nutrients can lead to competition between mycorrhizal fungi and other organisms for resources in the soil matrix. Gadgil & Gadgil (1971, 1975) suggested that ectomycorrhizal fungi of *Pinus radiata* reduced litter decomposition rates in part by competing with decomposer fungi for limiting resources. Herrera *et al.* (1978) found that ectomycorrhizal fungal hyphae were in direct contact with decomposing leaf material, thereby transporting P to the attached living root. Mosse & Phillips (1971) and St John *et al.* (1983a) found increased extramatrical VA mycorrhizal hyphae if organic compounds were added to the growth medium and Allen & MacMahon (1985) found a positive spatial correlation in the field between decomposing litter and external VA mycorrhizal fungal hyphae.

Interactions between prokaryotes and mycorrhizal fungi can be important for both organisms. Rose & Youngberg (1981) found that VA mycorrhizae and N-fixing actinomycetes increased growth more than either added alone (in this case, the prokaryote was *Frankia*, an actinomycete associate). Carpenter & Allen (1988) found the same interactive effects of VA mycorrhizal fungi and *Rhizobium* on the growth and survival of a semi-arid legume *Hedysarum borealis* in the field. Specifically, when either microorganism was added separately, neither promoted major alterations in growth or survival. However, when both *Rhizobium* and VA mycorrhizal fungi were added, growth, survival and flowering were all enhanced. Reports of interactions between P-solubilizing bacteria continue to suggest some form of interaction (R. Azcon-Aguilar *et al.*, 1986) and the interactions between mycorrhizal fungi and actinomycetes in forming and decomposing oxalates appear to be important in elemental cycling (particularly P, Fe and Ca) in both forest (Knutson *et al.*, 1980) and arid shrublands (Jurinak *et al.*, 1986).

In summary, mycorrhizal fungi are a complex group of organisms of many taxa including different fungal Divisions. Their only similarity is a convergent evolution toward a similar habitat and symbiosis type. They apparently compete with each other and with neighbors, can be defined in terms of niche dimensions as can higher organisms, and interact directly or indirectly with all members of their communities.

Plant community interactions

By altering the resource acquisition of the host plant, mycorrhizae affect the patterns and mechanisms of interactions between plants. That mycorrhizae can alter the structure and composition of plants within communities has been demonstrated for a variety of locations and communities. Janos (1981) reported that mycorrhizae increased both diversity and productivity of tropical tree communities. Newman *et al.* (1981) found that the abundance of mycorrhizae was closely correlated with the abundance of *Plantago lanceolata* in British grasslands. Hoffman & Mitchell (1986) suggested that the successful invasion of South Africa by *Acacia saligna* was associated with its mycorrhizal activity. In a careful microcosm experiment, Grime *et al.* (1987) determined that the addition of mycorrhizae increased plant diversity by raising the biomass of subordinate species.

In successional sites, mycorrhizae may be an important regulator of plant community composition. Hall (1979) and Powell (1980) noted that eroded sites in Australia and New Zealand, respectively, had low mycorrhizal activity and that the addition of mycorrhizae could result in increased vegetation. E. Allen (1984) suggested that mycorrhizae increase plant diversity in early successional communities by allowing non-mycotrophic species, or species with low mycorrhizal responses to exist where inoculum is low or not present and mycotrophic species to predominate where inoculum is high. The inoculum activity can be different across just a few centimeters (from no mycorrhizal inoculum to 100 spores per gram soil: M. Allen & MacMahon, 1985). Pendleton & Smith (1983) found a wide range in mycorrhizal fungal density including both mycorrhizal and nonmycorrhizal plants in the ruderal vegetation in Utah associated with disturbance.

Indeed, experimental results suggest that mycorrhizae may be as important as grazing (Grime *et al.*, 1987) or soil fertility (E. Allen & M. Allen, 1990) in determining plant community composition. However, the mechanisms of plant–plant interaction mediated by mycorrhizae are not well understood (Fitter, 1985b). Certainly, the simple assumption that there is a single action (e.g. improved P uptake via the fungal mycelium) for all plants belies the complexity of outcomes of plant–plant interactions that result from the presence or absence of mycorrhizae. To understand these, descriptions of the range of potential interactions between mycorrhizal fungi and host plants in controlled experiments need to be integrated with field observations of patterns resulting from mycorrhizal additions or removals. Unfortunately, few of these studies are in the literature, making generalizations difficult. Nevertheless, I will attempt to describe the characteristics that make the mycorrhizal association important to plant community functioning and then relate these to observed patterns of the effects of mycorrhizae on plant community structure.

Root distribution and its importance in resource acquisition has been documented since the early 1900s. Weaver & Albertson (1943) pointed out the importance of deep versus shallow roots in plant survival during the great drought of the 1930s in the Great Plains of North America. Root overlap and separation are known to be important aspects of niche dimension regulating the intensity of plant competition and plant coexistence in communities (e.g. Parrish & Bazzaz, 1976). Root distribution and activity are presumed to be the major determinants of soil resource activity and that distribution regulates the ability of plants to grow and survive (e.g. M. Allen, 1988b).

However, roots alone do not account for all, or even in some cases, the majority of nutrient uptake; they may serve primarily as a means to attach the mycorrhizal fungal hyphae to a plant. Ericoid mycorrhizal fungi account for virtually all of the N uptake of *Calluna* species in moorlands (e.g. Read, 1983). Achlorophyllous plants and orchids gain virtually all of their nutrients and energy via the fungal hyphae. Root systems of trees forming ectomycorrhizae have stunted absorption roots without adequate surface area for nutrient and water uptake (e.g. Frank, 1885; Hatch, 1937). Even VA mycorrhizae reduce the K_m of P uptake (Cress *et al.*, 1979) and can double the rates of water throughput of a plant with similar-sized roots (M. Allen, 1982; Hardie, 1985). The mycorrhiza as an entire unit must be considered in any understanding of soil resource gain by plants.

A mycorrhiza is a root plus its mycorrhizal fungus. The fungus acts, as does a root, by depleting resources and competing or cooperating with adjacent hyphae. In most cases, the fungal hypha incorporates the same forms of soil minerals as does a root. Mycorrhizal hyphae of the same compatibility group may fuse, creating a single mycelial network similar to the root grafting of plants. The resulting mycelial network may encompass several host plants, often of different species. We know little regarding the rates and extent of material flows through hyphae and independence of action of the various hyphal segments that make up that mycelium. This creates a series of key questions about the role of mycorrhizae in regulating the degree of competition or cooperation of plants in resource uptake. Thus, the degree of spatial and temporal separation among the various mycelial networks comprising the mycorrhizae in a community is important in describing the activity of mycorrhizae.

While mycorrhizae are part of the root system, the fungi and roots must coexist spatially only at the interfaces. In fact, since both organisms take up similar resource forms, a spatial overlap would simply increase the resource depletion and the two components of the mycorrhiza, root and hypha, would be competing. The fungus never occupies all of the root length. Most observations suggest that the mycorrhiza is associated with finer absorbing roots (e.g. Harley & Smith, 1983). These can be near the surface, as in tropical and temperate forests with segregated litter layers and little soil (e.g. Stark, 1972), or concentrate deep in arid sandy soils where the roots proliferate near the permanent water below the sand (Virginia *et al.*, 1986). They can be distributed throughout a deep soil profile such as the prairie soil of the Eastern Great Plains of North America (Zajicek *et al.*, 1986).

In addition to vertical spatial separation of roots to reduce competition, mycorrhizal fungi also may be vertically separated among plant species (Figure 5.6). Staffeldt & Vogt (1974) found that *Larrea tridentata* and the cacti in desert soils had shallow mycorrizhae (4–12 cm) whereas *Yucca elata* had mycorrhizae generally below 10 cm. Virginia *et al.* (1986) noted that the mycorrhizae of *Prosopis* sp. tended to be most active near the water table, down to 4 m deep. The potential for different spatial characteristics should be incorporated to understand the potential for mycorrhizal effects on plant community structure.

On a smaller scale, the fungi are rarely distributed directly where the mycelia compete with the roots for elemental absorption. In VA mycorrhizae the runner hypha extends down the root near the outer edge of the root hairs with the absorbing mycelial network composed of finer hyphae growing away from the root. This causes an increased depletion zone to develop beyond the root hairs of a nonmycorrhizal plant (Owusu-Bennoah & Wilde, 1979). The hyphae of other mycorrhizal types are higher fungi that act with greater independence by forming rhizomorphs that rapidly transport materials longer distances (Duddridge *et al.*, 1980) and thereby move greater distances into the soil, including several meters from a host. Both roots and hyphae apparently rapidly proliferate when organic matter is contacted (St John *et al.*, 1983a,b). When P resources are applied to soil patches, the mycorrhizal activity appears to increase but sometimes without concomitant root proliferation (Andrews *et al.*, unpublished data).

Mycorrhizae can segregate activity in time, reducing the competitive interactions. M. Allen *et al.* (1984a) found that *Agropyron smithii* and *Bouteloua gracilis*, co-occurring grasses in the American Great Plains, had different seasonal VA mycorrhizal activity corresponding to the C_3 and C_4 physiological activity of the plants and suggested that this niche differentiation acted to reduce competition. Similar patterns of VA mycorrhizal activity have been found in spring-sown versus summer-sown crops (e.g. Jakobsen & Nielsen, 1983).

A further factor that influences the functioning of mycorrhizae in communities is the physiological response of the various host plants to the mycorrhiza. As far back as the turn of the century Stahl (1900) recognized that different host species responded differently to mycorrhizal infection.

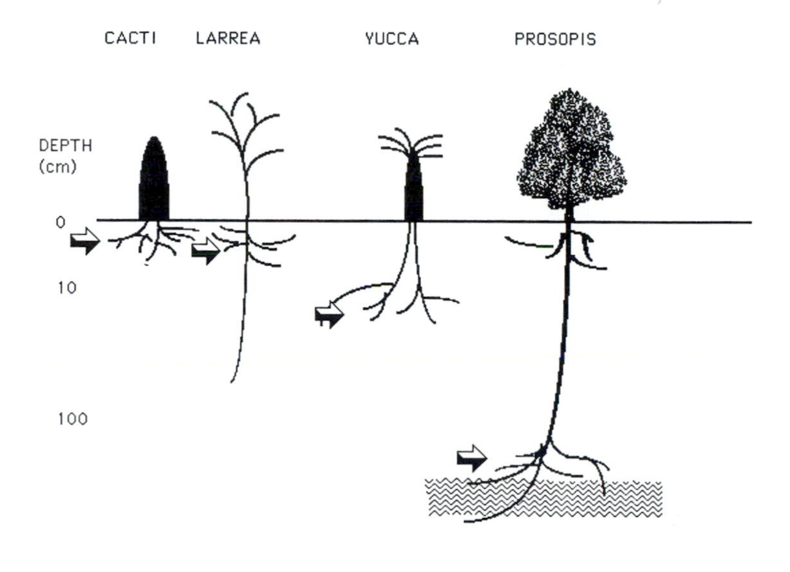

Figure 5.6. Segregation of VA mycorrhizal infection with depth among plants from the deserts of southwestern USA (from Staffeldt & Vogt, 1974, data used with permission from K. Vogt; and Virginia *et al.*, 1986).

He divided plant families into nonmycotropic (not forming mycorrhizae), facultatively mycotropic (in which the host can survive with or without infection), and obligately mycotropic (in which the host was thought to require the mycorrhiza to survive). Importantly, he provided no proof of 'obligate' infection, which led to a definitional dilemma that is still present.

Assessments of mycotrophy, the degree of 'feeding' of plants by mycorrhizal fungi, are generally based on considerations of the survival potential of the plant with respect to mycorrhizae in open, noncompetitive habitats (e.g. glasshouse experiments under ideal conditions). Nonmycotrophic plants 'never' form mycorrhizae (Stahl, 1900). However, mycorrhizal fungi can colonize even these nonmycotrophic plants (Tommerup, 1984; M. Allen *et al.*, 1989a). In some of these cases, the mycorrhizal fungus appears to act as a parasite, not a mutualist to the plant (M. Allen *et al.*, 1989a). M. Allen *et al.* (1989a) redefined nonmycotrophic plants as those that actively reject all mycorrhizal fungi. Facultatively mycotropic plants can survive with or without the mycorrhiza, and are probably the most common group of plants, although they certainly grade from those that show no response at all to the mycorrhiza (e.g. *Bromus tectorum*: E. Allen, 1984) to shrubs that demonstrate both enhanced growth and survival with mycorrhizae (e.g. *Atriplex canescens*, *Artemisia tridentata*, *Chrysothamnus nauseosus*) (Aldon, 1975; E. Allen, 1984; Lindsey, 1984). Finally, obligately mycotrophic plants are those that cannot survive without mycorrhizae. This category consists primarily of trees and shrubs that have never been observed to survive for long time periods without mycorrhizae or in which their survival following outplanting was dependent upon the addition of mycorrhizae (Baylis, 1975; Janos, 1980).

These categories are realistic only in the community sense. Virtually any plant can live under ideal conditions without mycorrhizae (in a glasshouse with high nutrient additions). Moreover, as Harley clearly pointed out, symbioses between microorganisms and plants are always somewhere along a gradient from parasitism to mutualism: always facultatively mycorrhizal. However, many plants do approach obligately mycorrhizal, particularly many long-lived trees and shrubs that are either in highly competitive environments (which makes nutrients limiting by being immobilized into living tissue) or in highly nutrient stressed environments where the fungi are essential for plant nutrient uptake (such as heathlands where nutrients are highly bound in the detritus) (E. Allen & M. Allen, 1990).

Different mycorrhizal types alter the structure and composition of the plant community but few experimental data have tested this hypothesis. Early attempts to establish ectomycorrhizal trees in grassland habitats (dominated by VA mycorrhizal plants) were not successful until the introduction of the ectomycorrhizal fungi. For example, in the American Great Plains (Goss, 1960) and the Ukraine (Schemakhanova, 1962), the establishment of shelterbelts to reduce wind erosion required the simultaneous introduction of the ectomycorrhizal fungal associates of the trees. I find no data for testing the mechanisms of interaction, whether competitive or open resources. In the Nebraska National Forest (*Pinus contorta* introduced into tallgrass prairie in the 1930s), few VA mycorrhizal grasses were found directly beneath the pines but the grasses predominated between the trees (Figure 5.7). Ectomycorrhizal and VA mycorrhizal plants are often interspersed in communities but nothing is known regarding their interactions. In some communities, e.g. the Pacific Northwest, the same fungi form arbutoid mycorrhizae with ericaceous shrubs and ectomycorrhizae with later-successional trees (Perry *et al.*, 1989), making competition among plants for mycorrhizal resources an interesting question.

Figure 5.7. The Nebraska National Forest showing the spatial segregation between the planted pines (*Pinus contorta*) with ectomycorrhizae and grasses (primarily big bluestem, *Andropogon hallii*) with VA mycorrhizae (M. Allen, unpublished observations).

Mycorrhizae alter the competitive outcomes among plants that have different physiological responses (different mycotropic status). Nicolson (1960) noted that in coastal sand dunes, the nonmycotropic annual *Salsola kali* was replaced by mycotrophic grasses during plant succession. Others observed the same pattern in successional areas in the High Plains of Western North America (e.g. Reeves *et al.*, 1979; Miller, 1979; E. Allen & M. Allen, 1980) and in the tropics (Janos, 1980). These data tend to lead to the hypothesis that mycorrhizal fungi increase the competitive ability of a host plant to the detriment of a nonmycotropic plant. Experiments in both the field and glasshouse have supplied a limited confirmation of this hypothesis. E. Allen & M. Allen (1984) demonstrated a reversal in competitive outcome between *Salsola kali* and two grasses, *Agropyron smithii* and *Bouteloua gracilis*, with versus without mycorrhizae. With the mycorrhizal fungi, the grasses improved their competitive ability, especially for water, a limiting resource in the habitats of origin. Without the mycorrhizal fungi, the annual was more competitive. These experiments have been repeated for a number of other pairs with similar results (e.g. E. Allen & M. Allen, 1990). Crowell & Boerner (1988) studied the effects of mycorrhizae on competition between annuals that were nonmycotrophic (*Brassica*) and VA mycorrhizal (*Ambrosia*). In this case, the nonmycorrhizal *Ambrosia* had virtually no growth without the endophyte with or without competition. Interestingly, the *Brassica* had a greater effect on mycorrhizal *Ambrosia* than did another *Ambrosia*.

The means whereby mycorrhizae enhance the growth, nutrient, or water status of one plant over another is not always clear but is presumed to relate to enhanced resource extraction by the mycorrhizal plant. The addition of VA mycorrhizal fungi in the field improved the ability of mycotrophic grasses to extract soil water when competing with annual weeds (E. Allen & M. Allen, 1986) but no differences due to mycorrhizae in nutrient composition were observed. Yocum (1983) found that the presence or absence of mycorrhizae altered the community composition when that community was a mix of mycotrophic and nonmycotrophic plants. He attributed the observed responses to enhanced P uptake.

Complicating the competition interpretation, however, was the finding that the VA mycorrhizal fungi also directly inhibited the growth and survival of nonmycotrophic annuals. Contradictory data suggested that nonmycotrophic plants (e.g. Brassicaceae, Chenopodiaceae) did or did not inhibit mycorrhizal formation of host plants. Iqbal & Qureshi (1976) found that *Brassica campestris* inhibited VA mycorrhizal formation in wheat plants. Ocampo *et al.* (1980) and Ocampo & Hayman (1981) noted that a wide variety of nonmycotrophic plants had no effect on VA mycorrhizal activity of a mycotrophic crop and even found some stimulation of VA mycorrhizal fungal activity when a field of a mycotrophic crop was preceded by nonmycotrophic plants. Schmidt & Reeves (1984) found that VA mycorrhizal fungi survived in the presence of *Salsola kali*, a nonmycotrophic annual. Alternatively, when no plants were present, in the same soil, the fungi could not survive living on organic matter alone. M. Allen & E. Allen (1990) further found that these fungi could grow and sporulate on carbon provided by nonmycotrophic plant species. E. Allen & M. Allen (1988) found that the addition of VA mycorrhizal fungi reduced the growth and survival of *S. kali* in the field and M. Allen *et al.* (1989a) demonstrated that the VA mycorrhizal fungi invaded seedlings of *S. kali* in the laboratory and even killed some seedlings. Whether similar patterns could be found between other incompatible plant–mycorrhizal fungus interactions is unknown. These responses indicate the complexity of interactions between mycorrhizal fungi and members of a host plant community.

Mycorrhizae also alter competitive outcomes among plants of relatively similar mycotrophic status. Crush (1974) and Hall (1978) demonstrated that mycorrhizae improved the competitive ability of legumes when grown with grasses. Fitter (1977) found that mycorrhizae reversed the competitive outcomes between two grasses, *Lolium perenne* and *Holcus lanatus*. Both studies attributed the results to increased responsiveness of some species (e.g. *Holcus*) to mycorrhizae compared with others (e.g. *Lolium*) among these facultatively mycotrophic species.

Mycorrhizae can be an important factor in structuring communities within a successional sere by making more resources available to some versus other plants. In a detailed double-labeling study in the field, Caldwell *et al.* (1985) found that a grass with higher rooting density (*Agropyron desertorum*) than another grass (*A. spicatum*) had greater total mycorrhizal activity (with little significant difference in proportion of root length occupied) and acquired a greater proportion of soil P in competition with a third species (*Artemisia tridentata*). These data supported the prior data, demonstrating the improved competitive ability of *Agropyron desertorum*.

In a summary analysis, E. Allen & M. Allen (1990) calculated crowding coefficients from a number of experiments. The resulting coefficients were similar to those calculated from fertilization experiments, and suggested that mycorrhizae are as important as nutrient concentrations in determining plant interactions. Grime *et al.* (1987) concluded that mycorrhizae were as important in regulating community structure as grazing, although for different reasons. Specifically, preferential grazing on the dominant plants increased evenness by increasing the resources available for the subordinate species. Mycorrhizae increased evenness by forming a common mycelial network and allocating resources to all infected plants. Few current competition studies incorporate mycorrhizae into their experimental design. These experiments are generally conducted in the field (where mycorrhizal inoculum is present) and compared with glasshouse studies that utilize sterile soils, where mycorrhizal fungi are absent. Oftentimes, the results of the field and glasshouse studies are reversed, highly suggestive of the importance of mycorrhizae (e.g. see the volume by Grace & Tilman, 1990, for several examples). Habitats where mycorrhizal fungi are absent do exist and are primarily highly disturbed areas (e.g. Mount St Helens: M. Allen, 1987a; strip mines with no topsoil: Schramm, 1966; E. Allen & M. Allen, 1980; eroded soils: Powell, 1980; and pesticide-treated soils: Menge & Grand, 1981). Incorporating this symbiosis into community and population studies is important for understanding the mechanisms of plant competition.

One mycelial network can encompass multiple plants, and fungal associations of multiple plants can, if they are compatible, anastomose (fuse) to form a single mycelium. Thus, the ability of any one fungus to colonize more than one individual plant, including plants of different species, means that a mycorrhizal association has the potential for initiating cooperative as well as competitive interactions between plants.

Woods & Brock (1964) suggested that forest trees can share resources via their mycorrhizal hyphae. They cut down one individual, labeled the stump with isotopes of Ca and P, and found the isotopes in several surrounding individuals. Chiarello *et al.* (1982) found a similar pattern in an annual, VA mycorrhizal community. These field studies demonstrated connections between plants, but neither could account for the pattern of nutrient distribution by plant size or distance.

A number of controlled experiments have validated the transport of P between plants via the mycorrhizal fungus. In perhaps the most detailed experiment, Whittingham & Read (1982) found transport of ^{32}P between plants interconnected by mycorrhizal hyphae but no transfer between a VA mycorrhizal plant and a nonmycotrophic plant. Francis *et al.* (1986) further found that this transfer improved growth of the receiver plant. However, Ritz & Newman (1984) suggested that the labeled P was primarily exchanged between two plants and no net transport had occurred. Ritz & Newman (1985) further suggested that the exchange between plants was a function of loss from one plant to the soil via exudates or dead roots and uptake by the mycorrhizae associated with the adjacent plant. The possibility of nutrient transport among plants in patchy environments has also been hypothesized (E. Allen & M. Allen, 1990). In this scenario, patches with high nutrients (as under a shrub) would be occupied by plants that are interconnected by mycorrhizal mycelia to plants existing in low nutrient patches. Excess nutrients might be transferred in such a case. This scenario has not yet been adequately tested. In a simulated sward, Grime *et al.* (1987) found that addition of mycorrhizae increased the diversity of the community by increasing the mass of the subordinate individuals. They suggested that the mycorrhizal fungi, by interconnecting several individuals, equalized resource allocations and allowed the 'less competitive' species to coexist. These data conflict with those of Caldwell *et al.* (1985), who demonstrated that even if the plants were interconnected, nutrients were preferentially acquired by one individual over the neighbor.

In ectomycorrhizal systems, the evidence is just as difficult to resolve. However, although transport has been observed (e.g. Brownlee *et al.*,

1983), Finlay & Read (1986) found that ^{32}P flow was unidirectional but they noted that the mycelial network picked up P from a single addition point and transported it throughout the mycelial network that included several individual plants of different species.

A series of glasshouse experiments has demonstrated that carbon also could be exchanged between plants via the fungal mutualist. Reid & Woods (1969), Hirrel & Gerdemann (1979), and Francis *et al.* (1986) found ^{14}C in acceptor plants interconnected by mycorrhizal hyphae to a donor plant that fixed $^{14}CO_2$. Read and colleagues, in a series of papers (reviewed in Read, 1984 and Read *et al.*, 1985) demonstrated that CO_2 fixed by one plant was transported from the plant to its mycorrhizal associate and could be found in the roots of an adjacent seedling connected to the same mycorrhizal fungus. They suggested that this flow was enough to enhance the survival capacity of the seedling under stress (e.g. Duddridge *et al.*, 1988). Finlay & Read (1986) further found that shade plants received more carbon from the fungus than plants in the sun that were photosynthesizing at a greater rate. The actual rates of transport in all studies were small and the significance to receiver plant survival awaits critical field studies for validation.

In summary, mycorrhizae regulate the composition and functioning of plant communities by regulating the resource allocation and growth characteristics of interacting plants. This regulation can take the form of transporting nutrients preferentially to one individual over another or of redistributing nutrients and carbon throughout the community. Finally, the ability of the symbiosis to exist where either the plant or the fungus alone does not, indicates the need to understand communities in the context of both fungal and plant components.

Animal community interactions

When ecologists debate the importance of fungi or animals, they generally discuss the relevance of the groups to plants. Animals graze on plants or disperse their seeds. Fungi decompose plant residue and form mycorrhizae that increase the plant's resource acquisition. Generally ignored are the interactions of animals and fungi and the effects of that interaction on the plant community. A community is composed of all of its components, animal, plant, fungal, or bacterial. Mycorrhizal associations interact with animals via both of the symbionts. Both animal and fungus depend on the plant for carbon. Thus, any grazing by animals reduces the carbon available for the fungus. Alternatively, the mycorrhizal fungus acquires resources for the plant that enable the plant to fix more carbon although not all that increase goes into increased plant biomass. Animals also feed directly on fungi, affecting the fungus both negatively and positively; negatively by direct consumption and positively by dispersing the diaspores of the fungus. Because of these conflicting interactions, the roles of animals in the community ecology of mycorrhizae might be the most integrative aspect of mycorrhizal ecology.

Grazing animals and mycorrhizal fungi depend on plants for energy. That requirement can be significant in the life of the plant. Estimates for carbon taken by above-ground animal grazers exceed 10% of the net primary production of a stand over a growing season and below-ground grazing can account for up to 25% of the production (e.g. Stanton *et al.*, 1981). Mycorrhizal fungi require between 10 and 30% of the net primary production (e.g. St John & Coleman, 1983), suggesting that the energy gained by one group radically affects the other (Figure 5.8).

Results from the few studies that have attempted to quantify these responses are highly contradictory. Bird *et al.* (1974) found that the reduction in nematode root parasites increased VA mycorrhizal activity. Alternatively, Stanton *et al.* (1981) found no significant change in mycorrhizae with the reduction in nematodes following nematicide treatment. Bethlenfalvay and colleagues (Bethlenfalvay & Dakessian, 1984; Bethlenfalvay *et al.*, 1985) found that heavy grazing by cattle reduced mycorrhizal activity and altered fungal species composition in a planted rangeland dominated by *Agropyron desertorum*. Reece & Bonham (1978) and Davidson & Christensen (1977) found that heavy grazing increased mycorrhizal activity in a shortgrass prairie. Wallace (1987a) noted that grazing by large ungulates did not alter mycorrhizal development but that trampling reduced mycorrhizal activity.

Several reasons may account for the lack of consistency in the results of experiments on mycorrhizae and grazing. The first revolves around compensatory response. Mycorrhizae not only require carbon but also stimulate the rates of photosynthesis. M. Allen (1980) found that VA mycorrhizal fungi increased photosynthesis by up to 80% in *Bouteloua gracilis* but appeared to utilize approximately 40% of that 80% increment resulting in a net carbon gain to the plant of about 30%. Harris & Paul (1987) noted that different plants demonstrated a range in net carbon change with mycorrhizae for different plants but, in general, a compensatory response was noted. The responses to grazing are highly controversial, ranging from those that suggest that grazing stimulates photosynthesis and/or growth (e.g. McNaughton, 1983) to those who suggest that there is seldom such a response (Belsky, 1986).

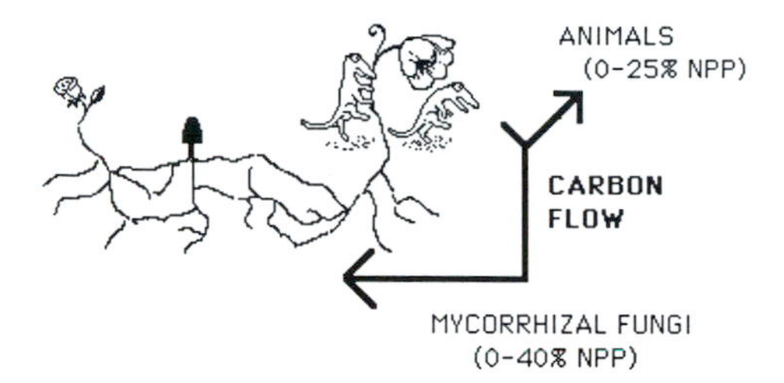

Figure 5.8. Competing carbon demands of animals and mycorrhizal fungi on a host plant. If one component depletes the energy resources of the plant, the other must suffer from an energy shortage.

A second alternative is that the loss of photosynthetic area by grazing must reduce mycorrhizal activity because of the resulting lower total carbon gain. Shading appears to reduce mycorrhizal activity in this manner (e.g. Johnson, 1976). However, if total light interception is not the limiting factor to growth, reduced leaf area might not reduce mycorrhizal carbon gain. For example, in arid conditions when water is limiting, the reduced leaf area might allow continued carbon gain by the plant whereas a larger leaf area would result in high plant resistance to water transport and closed stomates. Clipping of the leaves in two semi-arid bunchgrasses indicated no measurable differences in mycorrhizal activity (M. Allen *et al.*, 1989c).

The final alternative is simply that mycorrhizae are active at a smaller scale that does not allow accurate whole root system measurements. Carbon fixed by a plant is not simply transferred as a block to the entire root system. Allocations are made as a result of carbon demands and are usually associated with the sites of active resource acquisition. In periods of stress, mycorrhizal activity may be concentrated in small points in the soil and, thus, show no differences on a whole root system basis. This area needs more research, especially with insect grazers and pest outbreaks that may improve understanding of the interactions.

Unfortunately, I am aware of no data that have tested for the competitive carbon demands of mycorrhizal infections other than VA mycorrhizae versus those needed by grazers. This would be a fruitful area of research considering the importance of insect outbreaks on forests and on the wide array of insect grazers that exist at low levels in all forested and heathland communities.

Animals can directly inhibit mycorrhizal activity by feeding on the fungal hyphae and thereby reducing resource acquisition by the mycelial network. Invertebrates, in particular, gain much of their resources from fungi, including the mycorrhizal mycelium. Amoebae (Chakraborty *et al.*, 1985), nematodes (Ingham *et al.*, 1986), mites (Rabatin & Stinner, 1985; Moore *et al.*, 1986), and collembola (e.g. Warnock *et al.*, 1982; McGonigle & Fitter, 1988) all feed on mycorrhizal fungal hyphae. Fitter (1985a) and McGonigle & Fitter (1988) have suggested that this grazing may account for many field results that show a lack of response by plants to mycorrhizae. M. Allen *et al.* (1987) found that mycorrhizal activity was reduced in a cold desert during a series of wet years with high nematode numbers. Spores with puncture holes and internal bacteria were observed frequently during this time. Occasionally, the empty spores had nematodes inside them. This decreased mycorrhizal activity was correlated with decreasing shrub productivity and survival during what should have been favourable growth years.

Animals may stimulate as well as inhibit mycorrhizal fungi. In many areas, they may be essential to the dispersal of mycorrhizal fungi. MacMahon & Warner (1984) outlined three types of dispersal of mycorrhizal spores by animals. These included (1) ingestion of sporocarps as food, (2) movement of soil from areas with high spore densities to areas with low, and (3) phoresy, accidental transport of spores.

A wide range of animals are known to disperse mycorrhizal fungal inoculum from insects to large ungulates (see the extensive review by MacMahon & Warner, 1984). As early as 1922, this mode of fungal migration was noticed (Thaxter, 1922) and by 1939, the importance of mycorrhizal fungi in the diet of rodents was recognized (e.g. Diehl, 1939). Ingestion of spores or sporocarps and subsequent transport by animals has been demonstrated frequently from a wide variety of habitats (e.g. Emmons, 1982; MacMahon & Warner, 1984; Malajczuk *et al.*, 1987). Kotter & Farentinos (1984) and States (1984) particularly noted that tree squirrels ate sporocarps of hypogeous mycorrhizal fungi. Maser *et al.* (1978a) estimated that as much as 90% of the diet of some rodents was composed of sporocarps of mycorrhizal fungi. However, it is important that spores dispersed in this manner be tolerant of the gut tract environment. Several studies have demonstrated that spores can remain viable following excretion as faeces. These include studies of invertebrates (Ponder, 1980), rodents (e.g. Trappe & Maser, 1976; Rothwell & Holt, 1978; Warner *et al.*, 1987), and large ungulates (M. Allen, 1987a).

A wide variety of animals move soil that can result in the migration of mycorrhizal fungi. Gophers disturb soil as they search for roots and bulbs that form their food source. In undisturbed habitats, this results in a decrease in inoculum density (M. Allen *et al.*, 1984b; Koide & Mooney,

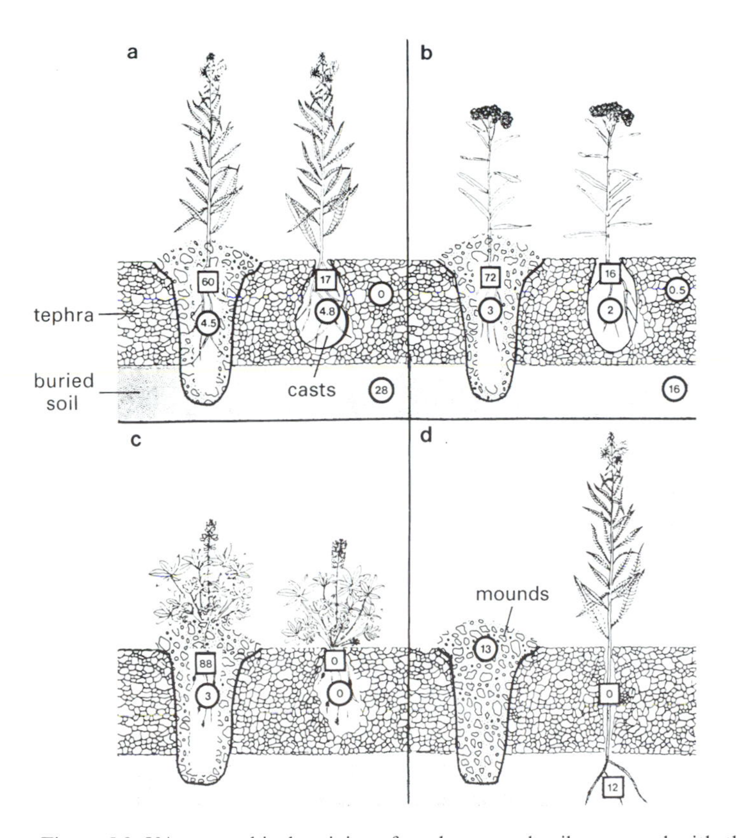

Figure 5.9. VA mycorrhizal activity of gopher-moved soil compared with the surrounding sterile ash material (tephra) deposited by the Mount St Helens eruption (from M. Allen, unpublished data). Shown is the activity in tephra, the old, buried topsoil, and in gopher mounds and casts including spore counts (○) and % root infection (□). The small letters represent four types of sites observed including (a) *Epilobium angustifolium* growing in a gopher mound and cast, (b) *Anaphalis margaritacea* growing on a mound and cast, (c) *Lupinus latifolius* growing on a mound and cast, and (d) where nothing was growing on a fresh mound with a nearby *E. angustifolium* growing in the tephra with its roots extending into the old soil.

1987). However, in some cases, such as a severe disturbance, inoculum can be increased. On the Mount St Helens eruption area, gophers and ants that survived the eruption moved the buried topsoil to the surface, resulting in patches of high inoculum concentrations and improved plant survival (Figure 5.9). In an arid shrub-grassland, harvester ants also concentrated mycorrhizal inoculum by lining their seed chambers and tunnels with roots containing a high density of mycorrhizal fungi. The resulting inoculum levels were more than an order of magnitude higher

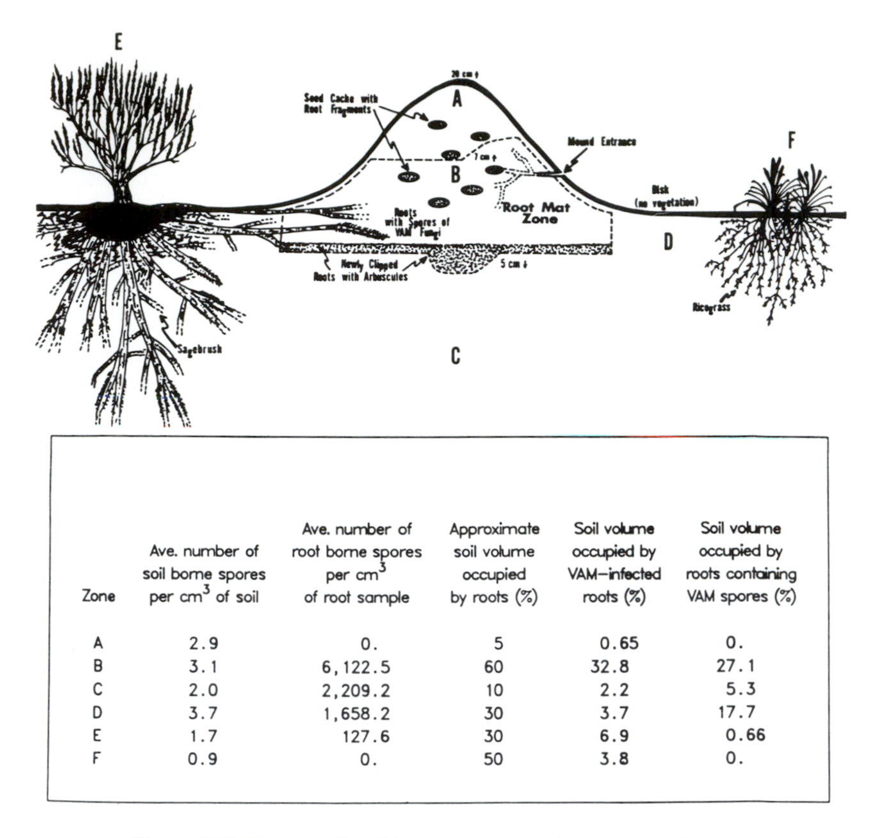

Zone	Ave. number of soil borne spores per cm^3 of soil	Ave. number of root borne spores per cm^3 of root sample	Approximate soil volume occupied by roots (%)	Soil volume occupied by VAM–infected roots (%)	Soil volume occupied by roots containing VAM spores (%)
A	2.9	0.	5	0.65	0.
B	3.1	6,122.5	60	32.8	27.1
C	2.0	2,209.2	10	2.2	5.3
D	3.7	1,658.2	30	3.7	17.7
E	1.7	127.6	30	6.9	0.66
F	0.9	0.	50	3.8	0.

Figure 5.10. Comparative VA mycorrhizal activity in an active harvester ant mound in a matrix of a surrounding undisturbed shrub-grassland community (from Friese & Allen, 1988).

within the ant mound compared with the surrounding undisturbed plant community (Figure 5.10).

Accidental accumulation and transport of mycorrhizal inoculum is also an important means of mycorrhizal fungal migration. Earthworms, millipedes, and wasps were observed to transport soil containing mycorrhizal inoculum (McIlveen & Cole, 1976). Wapiti on Mount St Helens fed on grasses in the meadows surrounding the disturbed volcanic area. As they fed, they tore the plants out and ingested the surface roots with viable inoculum. As these large ungulates moved across the volcanic debris (within 2 weeks of the eruption), they deposited the inoculum in these roots along with diaspores of plants. The plants subsequently sprouting from these faeces were mycorrhizal whereas nearby seedlings rooted in the tephra were nonmycorrhizal (M. Allen, 1987a). Gophers also carried inoculum, either ingested or in their surface hair, and moved it considerable distances (Allen & MacMahon, 1988). Grasshoppers have mycorrhizal spores on their surfaces. These spores can then be transported as far as the insect migrates or is blown (Warner *et al.*, 1987). Cicadas emerging from the soil were also observed to have surface-attached inoculum (M. Allen, unpublished data).

Mycorrhizal inoculum can be dispersed locally or over considerable distances by animals. Rodent and grasshopper dispersed inoculum was found all across a 75 ha disturbed strip-mine site (Warner *et al.*, 1987; M. Allen, 1988a) and animals apparently moved inoculum over several kilometres across the Mount St Helens volcanic eruption area (M. Allen, 1987a, 1988a). Truffles, a delicacy for any gourmet, are artificially transported over sometimes several hundred kilometres and inoculated (e.g. Chevalier & Grente (1977)) into habitats where the fungus did not exist.

In summary, mycorrhizal fungi and animals interact both positively and negatively with the possibility of considerable consequences for the community as a whole. The fungi and animals compete for plant carbon. The fungi also serve as food for animals resulting in both loss of fungal mass and the dispersal of the fungi. As plants, mycorrhizal fungi and animals all interact, the terms animal community, plant community and fungal community become simply components of the same community. None of these organisms are unaffected by the others and should never be treated as isolated units.

6

Ecosystem dynamics

Mycorrhizae potentially affect all aspects of terrestrial ecosystem functioning, from carbon allocation to nutrient immobilization. Mycorrhizal fungi may be the single largest consumer group of net primary production in many, if not most, terrestrial biomes. Moreover, by altering plant resource acquisition and plant production, mycorrhizae also dictate nutrient cycling rates and patterns. However, the literature on mycorrhizae and ecosystem dynamics is minimal compared with the literature on many other ecosystem processes such as grazing and N fixation.

Frank (1888) derived the first hypothesis of the role of mycorrhizae in ecosystem dynamics. He observed that mycorrhizal fungal hyphae predominated in the litter and humus layers of the forest floor and that the short tree roots covered with mycorrhizal fungi were inadequate for taking up needed nutrients, especially nitrogen. He suggested that mycorrhizal fungi utilize the humus to release and transfer nutrients, primarily N, from the forest floor to the trees. Melin and colleagues (see review in Melin, 1953) pursued this hypothesis, demonstrating that ectomycorrhizal fungi could transport labeled amino acids to the plant from the soil litter.

Stahl (1900) hypothesized that the most important mechanism whereby mycorrhizae regulate forests was by increasing water throughput, resulting in increased bulk flow of soil nutrients and increased deposition of nutrients in the roots. Hatch (1937) proposed that the mycorrhizal hyphae radiating out from tree roots directly took up and transported nutrients from the substrate into the plant, and Kramer & Wilbur (1949) demonstrated transfer of ^{32}P from soil to plant via mycorrhizal fungal hyphae. Thus, by 1950, the various means by which mycorrhizae could influence ecosystems were already recognized.

However, these hypotheses still concentrated on the individual plant–mycorrhizal fungus interaction and the resulting effects on plant

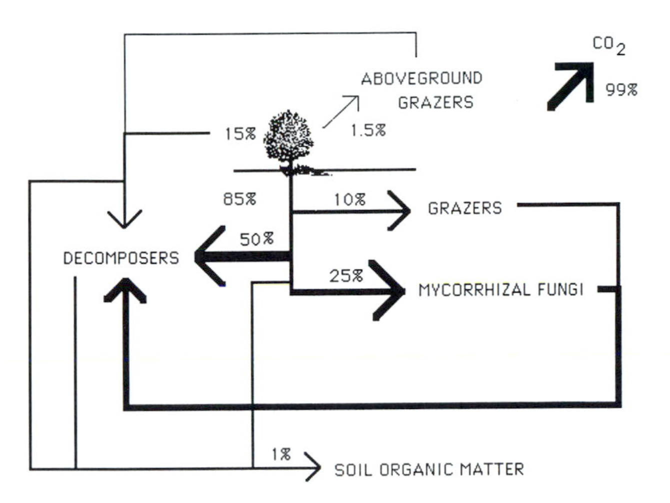

Figure 6.1. Allocation of carbon in most terrestrial ecosystems as a percentage of Net Primary Production (derived from Harley, 1971).

survival and productivity. Three pivotal papers, published in the 1960s, set the precedent for consideration of the roles of mycorrhizae in ecosystems. Woods & Brock (1964), by radioactive labeling of a cut stump of red maple, demonstrated transfer of ^{45}Ca and ^{32}P to other trees. They suggested that the mycelia of mycorrhizal fungi connected those trees and should be considered as a linking network for ecosystems. Went & Stark (1968) proposed that mycorrhizal fungi formed a large component of the soil fungi involved in rapid recycling of nutrients from decaying organic matter to plants. Odum (1969) suggested that in early successional or disturbed habitats, nutrients are primarily abiotic, ecosystems tend to have high entropy levels, and symbioses (primarily mycorrhizae) are not well developed. Later successional habitats have a high degree of symbiosis, organically bound nutrients and low entropy, suggesting that mycorrhizae are important in the successional process and ecosystem development.

Despite the exhaustive number of estimates of carbon and nutrient flows and above-ground trophic structure from the 1960s, Harley (1971) provided the first adequate ecosystem-level estimate of the importance of mycorrhizae in carbon allocation. More recent estimates (e.g. Fogel & Hunt, 1979, 1983; Vogt *et al.*, 1982; St John & Coleman, 1983; Harris & Paul, 1987) provide evidence that mycorrhizal fungi may comprise one of the largest carbon sinks for net primary production, equal to or greater than all animals together (Figure 6.1). However, the timing, spatial patterning, and ecosystem reaction to this allocation determine how nutrients cycle between soil and plant.

Several efforts have demonstrated that mycorrhizal fungi immobilize and transport a large proportion of the actively cycling nutrients in terrestrial ecosystems. Vogt *et al.* (1986) suggested that 18–58 % more N was added to the soil as ectomycorrhizae than was contributed by litterfall, and we have estimated that the N immobilized in VA mycorrhizal hyphae is equivalent to that in the soil solution (M. Allen, unpublished data).

In the ensuing sections, I will consider the roles of mycorrhizae as a regulating factor in ecosystem organization and functioning by assessing the contribution of mycorrhizae to carbon allocations, nutrient cycling and, finally, to succession.

Distribution of mycorrhizae in different biomes

By their role in the acquisition of nutrient resources, mycorrhizae are a critical component of nutrient cycling. Mosse (1975) has suggested that mycorrhizae should not be considered as a plant–fungus interaction but as a plant–fungus–soil partnership. Soils, mycorrhizal fungal types, and host plants change depending on their physical environment. Read (1983) has constructed a scenario of relationship among biomes, limiting resources, and mycorrhizal associations (Figure 6.2). In this scenario, the mycorrhizal types are fitted to physiology and structure, which regulates their ability to extract resources and regulate production both for improved plant growth and subsequent carbon for the fungus's own consumption. This scenario outlines a simplified approach to characterize the functioning of mycorrhizae on a distributional basis.

Beginning at the Equator (where our knowledge of mycorrhizal relationships is weakest), discernible trends in the distribution of mycorrhizal types are apparent. Within tropical rainforests, Meyer (1973) hypothesized that lower elevation forest trees are predominantly VA mycorrhizal whereas those trees found at higher elevations are mostly ectomycorrhizal. However, the few data collected indicate that the actual pattern may not be so simple. M. Gavito and E. Rincon (personal communication) have found only VA mycorrhizae in dry tropical forests at a range of elevations. Högberg (1982) found that the vast majority of trees in the Miambo woodlands in Tanzania were VA mycorrhizal but that in volume, ectomycorrhizal trees could dominate certain sites. Singer *et al.* (1979) reported that mycorrhizal fungi constituted the majority of sporocarps found in a tropical rain forest. More survey data of

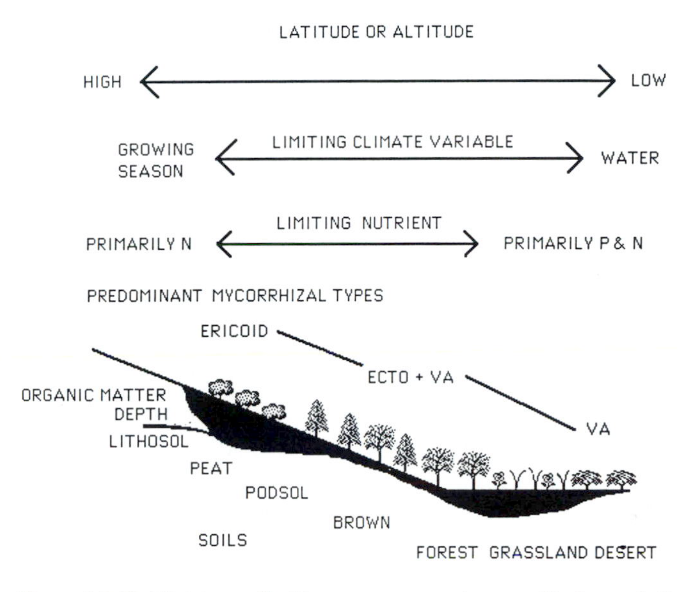

Figure 6.2. Habitat types, limiting resources, and mycorrhizal associations. A generalized overview of the relationships between soil types, limiting resources (both climatological and nutrient), plant types, and predominant mycorrhizal types (modified from Read, 1983).

mycorrhizal associates are needed in conjunction with current efforts in plant taxonomy. Tropical grasslands, like grasslands everywhere, are predominantly VA mycorrhizal. Deserts are dominated by VA mycorrhizal plants with occasional ectomycorrhizal trees found in oases (reviewed in Trappe, 1981).

Temperate vegetation has a rich mixture of mycorrhizal types. At low elevations, grasslands are dominated by VA mycorrhizae with occasional individuals of ectomycorrhizal trees. For example, in the North American tallgrass prairies, the widespread rich soils are dominated by grasses with VA mycorrhizae. These prairies are interspersed with rock outcroppings containing patches of ectomycorrhizal *Quercus macrocarpa* and bisected by river bottoms dominated by *Populus deltoides* and *Ulmus americana*, plants that form both VA and ectomycorrhizae. Lowland deciduous forests are a complex of VA- and ectomycorrhizal trees. The Bialowieza Forest in eastern Poland, probably the last of the great primeval forests of Europe (Falinski, 1986), contains a mix of ectomycorrhizal *Quercus*, *Fagus*, *Picea*, *Betula*, *Carpinus* and *Pinus*, VA mycorrhizal *Ulmus*, *Acer*, and *Fraxinus*, and many trees such as *Alnus* and *Tilia* that form both VA and ectomycorrhizae. The understorey consists of VA and ectomycorrhizal ferns, orchids such as *Goodyera* with their orchid mycorrhizae, *Monotropa* with its monotropoid mycorrhiza, and numerous VA mycorrhizal herbs. Similar complexes can be found in the rich diversity of the Eastern Deciduous Forest of the Eastern United States. For example, sandy soils tend to be dominated by ectomycorrhizal *Pinus* and the heavier soils with a mix of VA mycorrhizal *Liriodendron* and *Liquidambar*, ectomycorrhizal *Quercus* and trees that have both VA and ectomycorrhizae such as *Ulmus*, *Carya* and *Tilia*.

Chaparral and desert biomes tend to have mineral soils low in organic matter. In the deserts and semideserts of the Middle East or southwestern United States and northwestern Mexico, VA mycorrhizal plants predominate, although their seasonality and distribution often make the mycorrhizal fungi difficult to find (e.g. M. Allen, 1983; Virginia *et al.*, 1986). The chaparral may be the most interesting biome for study and yet is one of the least described. Ectomycorrhizal shrubs, such as *Quercus*, ericaceous shrubs (e.g. *Arbutus*, *Arctostaphylos*) with their arbutoid mycorrhizae, and VA mycorrhizal shrubs (e.g. *Artemisia*, *Salvia*, and *Rhus*), coexist and may be of equal sizes and dominance values in some stands (Figure 6.3). Orchids and achlorophyllous plants are also found in these habitats. As these vegetation types appear to have some convergence in form and physiology world-wide, the potential for comparative research among mycorrhizal associations and between geographical regions is virtually unlimited.

Similar patterns in mycorrhizal activity can be found at high altitudes and high latitudes. Boreal or subalpine forests tend to be dominated by ectomycorrhizal trees and ericoid mycorrhizal understorey with openings of VA mycorrhizal herbaceous meadows. Soils tend to be wetter and highly organic. In even more extreme circumstances of the tundra, ericoid mycorrhizae predominate (e.g. *Erica* and *Vaccinium*) with *Salix* and *Betula* (capable of forming both VA- and ectomycorrhizae) also being present. In more arid alpine habitats, VA mycorrhizal grasses and herbs often dominate (E. Allen *et al.*, 1987). Specialized 'dark-septate' fungi that exhibit some mycorrhizal characteristics are also found in many of these habitats (e.g. Haselwandter & Read, 1980).

Little is known about the mycorrhizae of polar regions. However, mosses and grasses, some of which can be infected with VA mycorrhizal fungi, are common. Island surveys at high latitudes tend to find numerous plants with ericoid and ectomycorrhizal fungi but no VA mycorrhizae. Ericoid, ectomycorrhizal, and dark-septate endophytic fungi were found on Devon Island, north of Hudson's Bay (C. Bledsoe, personal com-

Figure 6.3. A chaparral ecosystem that consists of plants with relative similar statures that contain differing mycorrhizal types. Shown are ectomycorrhizal oaks, arbutoid mycorrhizal manzanitas and VA mycorrhizal chamise.

munication) and Christie & Nicolson (1983) hypothesized that there are no mycorrhizal fungi in Antarctica based on the lack of VA mycorrhizae on plants of many of the islands surrounding the Antarctic mainland. Interestingly, I have found spores of *Gigaspora* in soil samples from Tierra del Fuego and of *Glomus* along with external and internal hyphae of endogonaceous fungi in the Toolik Lake region of the Alaskan North Slope (unpublished observations). Freckman *et al.* (1988) reported spores of an unidentified *Glomus*, all only known to exist as VA mycorrhizal fungi, from a dry valley in Antarctica and suggested that these spores were blown inland from the plant-colonized regions. These results indicate a need for additional observations especially comparing islands versus mainland sites and indicate either that the fungus may have been eliminated on small land areas or that there is a failure to colonize islands in these extreme conditions.

These surveys describe primarily the mycorrhizal status of 'natural' vegetation, i.e. that in a relatively undisturbed state (either by human activity or naturally). Disturbance and the complete alteration of vegetation by agriculture or forestry practices generally results not only in shifts in plant species composition, but also in the types of mycorrhizal associations and mycorrhizal fungal species. These anthropogenic effects will be described in the next chapter.

As indicated by this brief survey, we still know very little regarding the global distribution of mycorrhizae. Detailed regional surveys are limited even in temperate regions. The work in the tropics and Arctic, while increasing, is still well below that needed to make the most basic of generalizations. Surveys of mycorrhizal activity by ecologists as a part of tropical forest degradation, agricultural expansion in tropical grasslands, and energy development in the Arctic, could be done with minimal expertise and would provide a valuable basis for interpretation of those studies, and valuable information for mycorrhizal investigators.

Mycorrhizae and carbon dynamics

Detailed studies of carbon gain and above-ground allocations can be found for almost any terrestrial ecosystem. However, the below-ground system is divided into large black boxes labeled roots, soil microorganisms, and soil animals. Dead roots and leaf litter provide the carbon input that drives the other soil organisms and, in turn, they provide the nutrients required for production and nutrient resources (see, for example, the numerous IBP volumes devoted to trophic diagrams). The limit to such approaches is that it ignores the primary limits to production in most ecosystems, the basic composition of the below-ground biota, often the main carbon flows.

Light or CO_2 are rarely the dominant factors limiting whole-stand productivity (eg. Chapin *et al.*, 1987) without heavy fertilization and irrigation. Even in the Arctic, where light intensities are low, net primary production is generally regulated by low available nutrient concentrations or the limited time period when nutrients are available (e.g. Chapin *et al.*, 1988). Generally, the limiting factors to production are nutrients or water, resources provided by the roots and their symbionts. Several models of plant carbon allocation have demonstrated that carbon is allocated toward resource gain, i.e. if water is limiting, those roots growing into regions of higher soil moisture will preferentially gain carbon (Caldwell, 1976). Recent evidence suggests that carbon allocation to mycorrhizal fungi may function similarly. In a field experiment, when ^{32}P was placed in small patches within a root system, arbuscules became frequent, whereas these absorbing structures were absent in the adjacent, unfertilized mycorrhizal root segments (M. Andrews *et al.*, unpublished data).

Mycorrhizal roots and hyphae are often distributed where decomposing tissues are most prevalent. St John *et al.* (1983b) determined that mycorrhizal roots grew into resource patches (litter bags) more than into neighboring soil, and soil with organic matter had more hyphae than soil with no organic matter (St John *et al.*, 1983a). Their observations help explain field observations of higher concentrations of mycorrhizal hyphae in decomposing litter than in the surrounding soil matrix in both tropical (Herrera *et al.*, 1978) and arid (M. Allen & MacMahon, 1985) ecosystems. On a smaller scale, mycorrhizal fungi are often found sporulating in small patches of organic matter. Koske (1984), for example, found spores of *Glomus* inside decaying spores of *Gigaspora*.

Not only do mycorrhizae respond to organic matter, the fungi also provide much of that organic matter. Read (1984) suggested that mycorrhizal hyphae can be the largest microbial biomass component in many forests. *Hysterangium* mats were estimated to occupy up to 17% of the mineral soil volume in a Pacific Northwest *Pseudotsuga menziesii* site (Cromack *et al.*, 1979). In two estimates of the mycorrhizal contribution to forest carbon, Fogel & Hunt (1983) found that mycorrhizae comprised only 5% of the standing crop, but 50% of the organic matter throughput in a *Pseudotsuga menziesii* stand; Vogt *et al.* (1982) estimated that 15% of the net primary production in *Abies amabilis* forests was allocated to mycorrhizal fungi. Both estimates excluded the extramatrical hyphae due to the difficulty of differentiating mycorrhizal from saprophytic hyphae and living from dead hyphae. VA mycorrhizal fungi have up to 38 m of hyphae per cm^3 soil (E. Allen & M. Allen, 1986). Upon death of those mycorrhizal fungi, the hyphae contribute *c.* 1 mg of carbon in that cm^3 to decomposers. As much as 25% of the microbial biomass in the rhizosphere in a shrub-steppe can be comprised of VA mycorrhizal hyphae (E. Allen & M. Allen, 1986; M. Allen, unpublished data). By comparison, only 5% of the annual energy released by respiration is produced by animals in a tallgrass prairie, one of the most heavily consumed terrestrial ecosystems (e.g. Risser *et al.*, 1981). As the growth of the fungal hyphae can be rapid (up to several mm per day), and degradation rates are high, the carbon input can be considerable, especially to a local patch. Assuming that mycorrhizal fungi account for *c.* 15–20% of the annual net primary production, then carbon contribution to the soil could be significant.

Mycorrhizal hyphae create soil aggregates both by directly binding soil grains and possibly by producing polysaccharides that attach soil particles. Sutton & Sheppard (1976), Forster & Nicolson (1981) and Koske & Polson (1984) all noted that the external hyphae directly bound sand grains, increasing the stability of sandy beach soils. In some cases the mycorrhizal fungal hyphae appeared to produce a large fraction of carbon in those aggregates (e.g. Forster & Nicolson, 1981). These aggregates often are a major carbon input into systems with low organic matter.

The presence of mycorrhizal hyphae in soils adds carbon to the system, and also affects the decomposition of organic matter. This can occur either directly by decomposition of carbon substrates or indirectly by affecting the nutrient availability and thereby competing with decomposers for limiting resources such as N.

The hypothesis that mycorrhizal fungi can extract carbon from litter and transport it to plants (e.g. Falck & Falck, 1954), while certainly true for mycorrhizae of achlorophylous plants, may also occur to some extent in more common mycorrhizae. Certainly, many mycorrhizal fungi appear to be able to survive on organic substrates at least for a short time period. Warner & Mosse (1980) and Hepper (1983) suggested that VA mycorrhizal fungi could maintain a limited saprophytic growth phase. Ericoid mycorrhizal fungi are well known for their capacity to utilize organic nutrient sources (e.g. Read, 1983). Some ectomycorrhizal fungi also appear to have the capacity to break down even complex lignins and lignocellulose (Trojanowski *et al.*, 1984). The extent of this activity compared with that of general saprophytic fungi is relatively unknown but could be important especially where mycorrhizal fungi form a large part of the microbial biomass.

One hypothesis that has not been followed up adequately is that mycorrhizal fungi, by immobilizing N and P, reduce the availability of those nutrients so that the N or P limits the growth of saprophytic microbes and thereby retards decomposition (Gadgil & Gadgil, 1971,

1975). The implications of this activity are interesting, especially where the solution levels of N and P are low, either spatially or temporally, and mycorrhizal activity is high. For example, in decomposing logs, mycorrhizal activity can be high, increasing survival and seedlings especially during periods of drought (e.g. Harvey *et al.*, 1980; Christy *et al.*, 1982).

Mycorrhizae also alter ecosystem carbon dynamics by affecting carbon fixation by plants. However, no adequate stand-level models exist for mycorrhizal contributions to total fixation or fungal utilization. At the scales of individual plants, variability among species and conditions is high. M. Allen *et al.* (1984a) found that *Glomus fasciculatum* increased photosynthesis 80% in *Bouteloua gracilis* but only 40% in the co-occurring grass, *Agropyron smithii.* Two species of ectomycorrhizae, *Pisolithus tinctorius* and *Suillus granulatus*, increased photosynthesis of *Pinus contorta*, but that increment was age-dependent (Reid *et al.*, 1983). Even in the field, mycorrhizae increased photosynthesis by up to 80% in wheat (Trent *et al.*, 1989) and decreased stomatal resistance to CO_2 flux in native grasses by 30–50% (E. Allen & M. Allen, 1986).

By the same token, mycorrhizae alter the respiration of a site. As yet, there are no adequate stand-level estimates of the contribution of the mycorrhizal fungi to respired CO_2. Harris & Paul (1987) have estimated that only about 1% of the carbon fixed by a plant is respired by the mycorrhizal fungus and St John & Coleman (1983) have suggested that the figure could exceed 30% of the net primary productivity. However, the figures are complicated by the potential alterations of root respiration, soil water absorption of CO_2 and many other unknowns. It has been demonstrated that mycorrhizae increase soil CO_2 levels (Knight *et al.*, 1989) and certainly, based on the high biomass levels of mycorrhizal fungi in all but highly disturbed terrestrial sites, the mycorrhizal fungi must contribute a substantial amount to the total stand respiration.

Mycorrhizae not only increase the CO_2 exchange rates of plants with the atmosphere, but evidence from a wide number of studies also suggests that they transfer carbon between plants. The extent of transfer is debated and its importance at the ecosystem level is not understood. Reid & Woods (1969) demonstrated fluxes of labeled carbon from one plant to another via the mycorrhizal hyphae. Hirrell & Gerdemann (1979) found similar patterns in VA mycorrhizal systems. Read and colleagues (e.g. Read, 1984) suggested that C fluxes are especially frequent from mature plants to seedlings. Duddridge *et al.* (1988) described the patterns of that movement at the ultrastructural level by delineating the types and magnitudes of compounds moved. The magnitude of these transfers is in need of quantifying in a range of conditions (e.g. grazed plants, plants in shade).

In two of the most exciting recent papers, fixation of CO_2 by both VA and ectomycorrhizal fungi was clearly demonstrated. The VA mycorrhizal fungus fixed enough carbon to allow the mycelium to expand from initial germ tubes until a host had been encountered (Becard & Piche, 1989). The ectomycorrhizal fungus *Paxillus* also fixed CO_2 (Lapeyrie, 1988) which would enhance its growth during periods when active plant transport would be low. The ability of fungi to fix CO_2 has been known for several years (Niederpruem, 1965), although the rates for natural ecosystems and under a wide range of conditions are not known. Nevertheless, these rates could be extremely important to understanding carbon fluxes in ecosystems and certainly are important to the localized soil gas composition and subsequent effects on nutrient cycling.

Mycorrhizae are only one component of ecosystems, albeit an important one. As discussed in Chapter 5, fungus and root serve both as food resources and to alter production. However, unlike grazers where the evidence for compensatory response is still highly controversial (Belsky, 1986), mycorrhizal fungi almost always alter the plant in such a manner that the net carbon gain is higher with mycorrhizae than without them. The extent of that gain is not known. Virtually all ecosystem-level carbon models incorporate carbon movement from plant to consumers. Recent models even incorporate loss to below-ground consumers such as nematodes. Most efforts now also include a flow to the infamous soil microbial black box. However, mycorrhizal fungal hyphae have a direct pipeline to the host and transfer nutrients directly to that host. Thus, functionally, they act more as a root system extension than as part of the 'microbial biomass' (Figure 6.4). Detailed field estimates and models of the functional importance of mycorrhizae in carbon balance are clearly needed. Only when these are included can ecosystem models represent predominant flow patterns.

Mycorrhizae and nitrogen cycling

Frank first suggested that mycorrhizae are important in the uptake of N from forest floors. Melin and colleagues (Melin, 1953; Melin & Nilsson, 1953) demonstrated that, at the individual seedling level, mycorrhizae enhance the uptake of N and, in many cases, transport organic N from substrate to plant. Read and colleagues have documented the importance of ericoid and ectomycorrhizae in transferring organic N

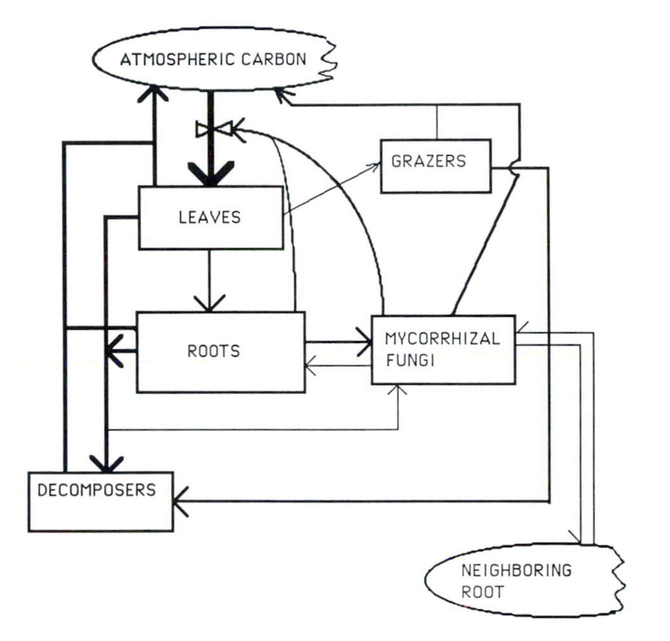

Figure 6.4. A conceptual model of the roles of mycorrhizae in ecosystem carbon dynamics. In this scenario, only roots and mycorrhizae have the capacity to stimulate a compensatory response to the carbon lost to them. Mycorrhizae also have the capacity to utilize decomposing carbon and transport carbon between plants, although the rates in nature are unknown (see text).

to the host plant (e.g. Stribley & Read, 1974; Bajwa & Read, 1986; Abuzinadah *et al.*, 1986; Abuzinadah & Read, 1986). Thus, since the earliest studies of mycorrhizae to the present, the role of mycorrhizae in N cycling has occupied a central focus in mycorrhizal research efforts.

Mycorrhizae can affect N cycling via several means. As they are a large biomass component, both the fungus and fine roots will immobilize substantial quantities of N in producing their own growth. Mycorrhizal hyphae have the capacity to extract N and transport it from soil to plant because of the enhanced absorptive surface area. In addition, mycorrhizal fungi contain enzymes that break down organic N and contain N reductase for altering forms of N in soil. Finally, mycorrhizae may act to increase N-fixation rates of N-fixing plants by alleviating other stresses imposed on those plants.

The range of estimates of N immobilized by mycorrhizae is extreme, from 15 g per m^2 or 1 % of the soil N standing crop in a cold desert (M. Allen, unpublished data), to 240 g per m^2 or 7 % of the total N in an *Abies amabilis* forest (Vogt *et al.*, 1982), to 1560 g per m^2 or 2% in mycorrhizae (as part of the mantle of the mycorrhiza) plus another 2% of the N in fungal (saprophytic plus fungal) biomass in another coniferous forest (Fogel & Hunt, 1979). However, as discussed in the carbon section, the turnover rates of these fungi are much higher than of most other materials. Vogt *et al.* (1982) estimated that the fungal component of the mycorrhizae contributed about 27 kg per ha per yr (27%) of the N turnover and the fine root + mycorrhiza another 60 kg per ha per yr (60%) of N turnover in the 23-year-old stand. They reported relatively similar values in a 180-year-old stand. Fogel & Hunt (1983) estimated that the return of N to the soil by mycorrhizae comprised about 83–87% of the tree return. The estimate by Vogt *et al.* (1982) does not include the extramatrical hyphae, a large and rapidly cycling biomass component, and utilizes only fine root mycorrhiza and fungal reproductive structures in their estimates. This is due to the impossibility (at present) of differentiating ectomycorrhizal hyphae from saprophytic hyphae and the inability to calculate individual hypha turnover. Hyphal production and turnover in the cold desert can be rapid; microbial biomass can double in 2 days in response to precipitation and then decline by 75% within another 2 days (M. Allen, unpublished data). This would suggest that the N throughput via mycorrhizal fungi can be much higher than the 1–2% bound at any one time. This limited number of estimates not only suggests that this pool is important but also indicates a lack of data to help us understand rates and amounts. A solid, quantitative modeling effort is needed to understand better the role of mycorrhizae in N immobilization and release.

Mycorrhizae also transport N along the fungal hyphae in a manner similar to P although the relative importance of this pathway is dependent on the forms of N in the soil. Nitrate is relatively mobile and much of it is transported in the soil solution by mass flow. As mycorrhizae can increase the water flow through plants (up to 100% during periods of drought, e.g. M. Allen, 1982), there is a potential for increased nitrate migration to roots for uptake. Although mycorrhizal fungi show some nitrate reductase activity, this appears to be low and the importance of mycorrhizae in absorption of nitrate, in general, is considered to be of lesser importance (e.g. Ho & Trappe, 1975; Bowen & Smith, 1981; Oliver *et al.*, 1983). Mycorrhizal fungi do have the ability to utilize and transport organic N, particularly as specific amino acids, to the host. Melin & Nilsson (1953) demonstrated transport of glutamic acid to pine seedlings via *Boletus* (*Suillus*) *variegatus*, an ectomycorrhizal fungus. Bajwa & Read

(1986) clearly demonstrated that, although both ecto- and ericoid mycorrhizae transport amino acids to the host, the ectomycorrhizal fungi tend to be more selective. The ability of ericoid mycorrhizae to utilize organic N is especially important. These associations are found predominantly in highly acidic, organic soils in which N uptake by plants is virtually nonexistent without mycorrhizae (e.g. Haselwandter & Read, 1980; Read & Bajwa, 1985).

In many natural ecosystems a large fraction of the available N in the soil is ammonium, not nitrate. Mycorrhizal fungi readily transport NH_4^+ from soil to plant, be they VA mycorrhizae (Ames *et al.*, 1983) or ectomycorrhizae (e.g. Bledsoe & Rygiewicz, 1986; Martin *et al.*, 1986). This may be especially important when the N is distributed in discrete patches. For example, Becker (1983) found that ectomycorrhizae of *Shorea* seedlings penetrated subterranean termite nests in a Malaysian rainforest. In a cold-desert shrubland, termites and ants were observed to occupy the same mounds (C. Friese and T. Crist, personal observations) and heavily infected VA mycorrhizal roots were observed penetrating those mounds (Friese & Allen, 1988). Termite nests represent large pools of N in many systems (e.g. Salick *et al.*, 1983) and the ability of plants to extract that N via the mycorrhizal hyphae could be an important pathway for N incorporation into actively cycling fractions.

Mycorrhizal associations also can enhance N gain in ecosystems by increasing the N-fixation rates of plant–bacterial N-fixing associations. Increases in the rates of N fixation with mycorrhizae have been observed in legumes (e.g. Hayman, 1987), actinorhizal associations (e.g. Rose & Youngberg, 1981; Gardner *et al.*, 1984), and in free-living associations (e.g. Pacovsky *et al.*, 1985). The means by which mycorrhizal associations increase N fixation rates is variable, but, within the physiological responses of a single plant to mycorrhizae, is dependent upon such factors as plant P concentrations, photosynthetic rates, and hormones (e.g. Puppi, 1983). At the ecosystem level presumably much increased fixed N becomes incorporated into the actively cycling N pool, especially in the soil. However, several qualifications should be placed on this assumption. By increasing plant growth, mycorrhizae also increase the N requirement of a plant and, if that N is not deposited where fixed it may not increase in the local soil. Carpenter & Allen (1988), for example, found that mycorrhizae plus *Rhizobium* increased the growth and seed production of *Hedysarum boreale* in the field. However, the N requirements apparently were not met by fixation alone as there was still a net depletion in soil N. Much of the plant N was apparently incorporated into the seeds or litter blown away from the parent plant. Thus, the spatial and temporal characteristics of mycorrhizae and N-fixing activity need to be understood before generalizations regarding the ability of mycorrhizae and N fixation to increase soil N can be validated for any site.

One means whereby mycorrhizal-enhanced N fixation may increase the N incorporated into the substrate may be via the hyphal connections between plants. Recently, several investigators demonstrated that N fixed in association with one plant can be transported to an adjacent, non-fixing plant via the mycorrhizal fungal mycelium. In crop systems, N fixed by soybeans was transported to maize via the VA mycorrhizae and significantly increased the growth and N status of the maize plants (Van Kessel *et al.*, 1985). Similar responses have been hypothesized between actinorhizal alder and adjacent plants via connecting ectomycorrhizae. These interconnections may be of critical importance to N distribution in ecosystems and need further research.

Mycorrhizae and phosphorus cycling

Mycorrhizae have long been known to affect the P nutrition of host plants. Phosphate, the major form of P available for uptake by plants, is relatively insoluble in the soil solution and therefore, is not readily transported by mass flow (Nye & Tinker, 1977). Thus, as mycorrhizal hyphae explore the bulk soil beyond the root hairs, additional P is taken up by the hyphae and transported to the host. This mechanism of activity has been thoroughly discussed in the literature dealing with the physiology of mycorrhizae (e.g. Hatch, 1936; Kramer & Wilbur, 1949; Sanders & Tinker, 1971; Hattingh *et al.*, 1973; Pearson & Read, 1973; Hadley, 1985; Harley & Smith, 1983; Safir, 1987). In understanding ecosystem P dynamics, of concern are the effects of mycorrhizae on the spatial dispersion of P, the potential roles of mycorrhizal fungi in the weathering process, and the total proportion of the P immobilized and mineralized.

Mycorrhizal hyphae, by growing into the soil matrix, can gain access to bulk soil P beyond the depletion zones created by the plant roots. This expanded depletion zone has been demonstrated (Owusu-Bennoah & Wild, 1979), and uptake by mycorrhizal hyphae has been demonstrated from as far away as several centimeters (e.g. Rhodes & Gerdemann, 1975; Read, 1984). In some cases, the depletion zone results in a change in the overall dispersion pattern of P at a site. The soil P concentrations under an *Artemisia tridentata* shrub increased with time in a successional shrub desert but declined in the associated interspace regions occupied by VA mycorrhizae (M. Allen, unpublished data). Mycorrhizae may be involved in the development of these 'islands of fertility' that characterize arid regions (M. Allen, 1988b; Skujins & Allen, 1986). Chapin *et al.* (1987) also noted that the ectomycorrhizal hyphae of the black spruce appeared to permeate the moss layer in Alaskan taiga forests and transported P to the tree. Upon application of fungicides to the mossy surface, P appeared to move down through the profile or remain in the mosses.

Mycorrhizal fungi can also increase the uptake of mineralized P by occupying the microsites of active decomposition and, possibly, by being involved in the degradation of litter (see N section). Herrera *et al.* (1978) found mycorrhizal hyphae on a decomposing leaf and rapid transport of ^{32}P from that leaf to a host plant via mycorrhizal hyphae. Wood *et al.* (1984) found that little of the P mineralized in an undisturbed forest floor moved through the soil profile and suggested that the mycorrhizal fungi rapidly recycle P into the host trees. With the disruption of the forest floor and the intact mycorrhizal mycelium, P rapidly moved into the soil and was transported out of the system. Allen & MacMahon (1985) found high spatial correlations among decomposer activity and labile organic matter with VA mycorrhizal hyphae.

By taking up P via the hyphae, mycorrhizal fungi also determine P allocation among plants interconnected by a mycelial network in a soil matrix. Finlay & Read (1986) found that P could be taken up by a single hypha and transported throughout a mycelial network that incorporated several plants. The relative allocation of P within the mycelium and then between plants may depend upon the photosynthetic activity of the host and the ability of the plants to compete for mycorrhizae. Reid and colleagues (e.g. Reid *et al.*, 1983) have suggested that a mycorrhizal plant has more resources, therefore it photosynthesizes more, therefore it gains more P as a result of the increased P sink. Caldwell *et al.* (1985) noted that plants with a greater rooting density and greater mycorrhizal fungal density gained more labeled P from the interspace than neighboring plants. This could result in the redistribution of P in the soil, creating habitats where resources are distributed in discrete patches (e.g. E. Allen & M. Allen, 1990).

The potential access by mycorrhizal fungi to P resources other than soil solution P has been highly controversial. Most workers have suggested that mycorrhizae mostly merely increase the surface area for P uptake (e.g. Sanders & Tinker, 1971; Safir, 1987). However, three mechanisms of mycorrhizal activity have been proposed that contribute to weathering of soil P as well as simple transport to the host plant. These are (1) the interaction of mycorrhizal fungi and P-solubilizing bacteria, (2) the production of phosphatases by the mycorrhizal fungus, and (3) the production by the mycorrhizal hyphae of organic acids that mineralize P.

Several recent studies have suggested that mycorrhizal hyphae may have the capacity to alter the weathering rates of soil P and increase the total P cycling in ecosystems. Owusu-Bennoah & Wild (1980) noted that with mycorrhizae, the unavailable soil P pool declined through time, not the plant available pool. Bolan *et al.* (1984) noted that the ^{32}P techniques previously used to determine the P sources were flawed in that simple ^{32}P added continually exchanges with other fractions and that some of those may not necessarily be available to plants. They amended soils with iron hydroxide followed with ^{32}P labeling and found that mycorrhizal plants had access to the introduced P source and that it was not available to nonmycorrhizal plants. Duce (1987) found further that it was primarily the bound inorganic P pool and not the organic P pools or the labile inorganic P that decreased with the increased plant growth associated with VA mycorrhizae, suggesting increased use of the bound inorganic P fraction. He further reported no significant increase in rhizosphere phosphatase activity associated with the mycorrhizae. These data suggested that mycorrhizae also affected the weathering of soil P.

For the first mechanism of soil weathering, Azcon *et al.* (1976) proposed that bacteria associated with mycorrhizal plants could improve the P acquisition of added rock phosphate in alkaline soils. In this scenario, the bacteria solubilized the P, and the P was then transported to the plant via the mycorrhizal fungus. This hypothesis, while not widely discussed, has not been rejected and appears to be important in some soil conditions (e.g. R. Azcon-Aguilar and J. Barea, unpublished data).

The second mechanism, the production of phosphatases by mycorrhizal fungal hyphae, has been widely demonstrated. Alexander & Hardie (1981) found extensive surface acid phosphatase activity of ectomycorrhizae in spruce from serpentine soils, and Bartlett & Lewis (1973) noted phosphatase activity associated with beech roots. Ho & Zak (1979) found that different ectomycorrhizal fungi had significantly different acid phosphatase activities that correlated with the soil characteristics wherein they were isolated. Antibus *et al.* (1986) and Kroehler *et al.* (1988) have published extensively on the components in artificial growth media that regulate acid phosphatases of ectomycorrhizal fungi: pH, temperature, and inorganic and organic P concentrations. Allen *et al.* (1981a) found no significant acid phosphatase activity associated with VA mycorrhizal fungal hyphae but observed increased alkaline activity with mycorrhizae.

Gianinazzi-Pearson & Gianinazzi (1983) noted that alkaline phosphatase activity was especially important in VA mycorrhizae. Neil (1973) found that many nonmycotrophic, weedy plant species demonstrated higher acid phosphatase activity than mycotrophic species in pot cultures grown in sterile soil. Dodd *et al.* (1987) noted a similar pattern but that the phosphatase activity increased upon infection in the mycotrophic plants when infected by VA mycorrhizal fungi.

For the third mechanism, Graustein *et al.* (1977) noted that some ectomycorrhizal fungi (especially, *Hysterangium*) produced calcium oxalate in relatively high concentrations. They further suggested that the production of this organic acid would increase the weathering rates of soils, increasing the cycling rates of important cations, especially Ca, Fe and Al (Graustein *et al.*, 1977; Cromack *et al.*, 1979). Jurinak *et al.* (1986) found crystals of oxalates associated with VA mycorrhizal hyphae. They further found that high CO_2 enhanced P weathering from clay soils of semi-arid habitats and completed a series of thermodynamic models that suggested that oxalates preferentially bound Ca, Fe and Al to phosphates. They proposed that mycorrhizae could enhance the availability of soil P by weathering P from the clay matrix and maintaining the solution P by binding Ca with the secreted oxalates (Figure 6.5). The oxalates would then be degraded by actinomycetes that again would enhance soil CO_2 and weathering. Knight *et al.* (1989) demonstrated that VA mycorrhizae could enhance soil CO_2 and Duce (1987) found increased available P through time associated with mycorrhizae. More recently, Koslowski & Boerner (1989) reported that inoculation by VA mycorrhizal fungi reduced the effects of soluble Al, particularly with respect to immobilizing P. These data suggest that mycorrhizae have the capacity to enhance not only the transport of P to a plant from the soil solution, but also to enhance the weathering rates of P from the bound, inorganic P pool and to reduce the rebinding capabilities of Ca and Al thereby increasing the actively cycling P in the ecosystem.

The total mycorrhizal immobilization of P, as with N, appears to be important over short time periods, and throughput via the mycorrhizae can be very high. Several studies have demonstrated that mycorrhizae can provide a large proportion, if not the majority, of the P taken up by the plant over a growing season. To transfer that increment, a substantial fraction of the biologically active P should be in the fungal mycelium. Few data exist to validate or refute this hypothesis. Vogt *et al.* (1982) estimated that 5% of the P in living biomass was in the reproductive structures of the mycorrhizal fungi and another 16% in the fine root mycorrhizal

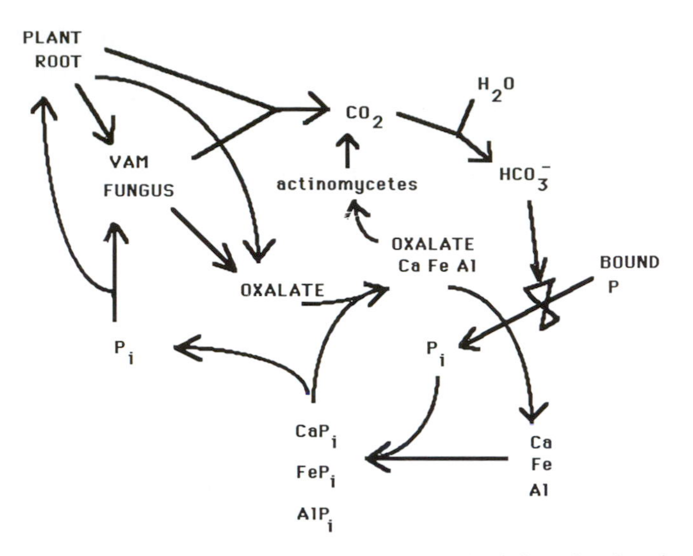

Figure 6.5. The role of mycorrhizal hyphae and hyphal-produced oxalates in the release of bound inorganic P, subsequent availability via cation binding, and uptake of that available P (see text for details, derived from Jurinak *et al.*, 1986).

sheath in a 23-year-old stand of *Abies amabilis*. The mycorrhizae of this stand also contributed 14 and 9 times the P turned over compared with litterfall in the 23- and 180-year-old stands, respectively. Fogel & Hunt (1983) estimated that 36% of the plant P was in the mycorrhizae and that mycorrhizae returned 51% of the annual P on a stand basis. Adequate estimates from VA mycorrhizal ecosystems are nonexistent.

In summary, mycorrhizae are clearly critical to understanding P cycling on a stand basis. The fungi alter the weathering of P from the bound pools, transport a large fraction of the total P moved from soil to plant, and immobilize and turn over perhaps the majority of stand P.

Mycorrhizae and other resources

In addition to N and P, mycorrhizal fungi take up and translocate almost any element required for plant growth, and mycorrhizae also enhance the transport of water from soil to plant (see Chapter 4 for details). Again, little is known regarding the relative mechanisms and rates of immobilization and transport at the ecosystem scale.

The effects of mycorrhizae on water flow through the soil–plant–atmosphere continuum have seldom been quantified and never calculated on a stand basis. However, it could be substantial. In southwestern

Wyoming the water relations of paired plots of mycorrhizal and nonmycorrhizal grasses were compared (E. Allen & M. Allen, 1986). In these cases, stomatal resistances to water fluxes were reduced by 40% in soils of similar moisture status. Under the high temperatures and low moisture conditions of the site, transpiration rates can be extremely high. There was no significant difference in the cover of mycorrhizal and nonmycorrhizal plants at the site. At this site, up to 32 ml per m^2 per hour was transpired in the mycorrhizal plots versus 14 ml per m^2 per hour in the nonmycorrhizal plots during peak transpiration periods (M. Allen and E. Allen, unpublished data). However, this does not represent the entire day. Mycorrhizal *Bouteloua gracilis* has been shown to shut its stomata more rapidly in response to low light or cloud flecks than nonmycorrhizal plants, thereby conserving water when light energy harvest was reduced (M. Allen *et al.*, 1981b). Also, the mycorrhizal plants were competing with nonmycorrhizal annuals and the increased water gain by the mycorrhizal plants enhanced its ability to survive that competition (E. Allen & M. Allen, 1986). In this case, the relative transpirational differences occurred during a short drought in an otherwise wet year and the effects of the enhanced transpiration (via the mycorrhizal plants) on soil moisture status were not large enough to detect. During normal precipitation years, the differences could be extremely important to the plant and soil water balance.

In a more recent study, Trent *et al.* (1989) fumigated a field plot and evaluated the resulting effects of mycorrhizae on plant production, grain filling, photosynthesis, water relations, and macro- and micronutrient concentrations. They reported that the major differences between treatments were an improved water relations, photosynthesis, and grain filling. No nutritive differences could be found. Interestingly, the changes in water relations included both increased conductance with lower leaf water stress (higher leaf water potentials) which would indicate that the plants had access to additional water sources (e.g. deeper roots or direct hyphal transport). This would indicate an altered soil water balance although specific changes were not tested.

Mycorrhizae regulate not only the uptake, but also the relative abundance of available and transportable nutrients. For example, siderophores are formed by mycorrhizal fungi that enable the fungus to take up Fe from solutions as low as 10^{-54} M (Szaniszlo *et al.*, 1981). The ability of mycorrhizal hyphae to weather and concentrate Ca and other cations (e.g. Cromack *et al.*, 1979; Lapeyrie & Bruchet, 1986) can enhance the total quantity of those elements in the biologically-active fractions.

Few quantifications of the roles of mycorrhizae in the accumulation and throughput of other elements are available. However, several ecosystem studies suggest that they vary in importance by element (e.g. Sollins *et al.*, 1980; Vogt *et al.*, 1981, 1982; Fogel & Hunt, 1983). More data are needed.

In an interesting twist on the roles of mycorrhizae in nutrient cycling, Boerner (1986) looked at the differences in seasonal nutrient dynamics and nutrient resorption of two perennial understorey VA mycorrhizal herbs (*Geranium maculatum* and *Polygonatum pubescens*) in differing environments. He found an inverse relationship between nutrient absorption efficiency and nutrient use efficiency. When the plants were heavily mycorrhizal (low soil nutrients), uptake efficiency was high whereas when mycorrhizal infection was low (high soil nutrients), nutrient use efficiency and resorption was high. No similar response was found in the trees in the high light environments. This would suggest that mycorrhizae could be important not only in the uptake and cycling rates, but also in how efficiently the plants absorb and utilize the nutrients absorbed.

Mycorrhizae and patch dynamics

Ecosystems should not be viewed as landscapes containing plants linked by a single mycelial network. A unit of land contains several vegetation units, whether a monocropped field with weed patches within, or a native meadow surrounded by a forest. That land unit also contains multiple mycorrhizal fungal species, inoculum densities, and often mycorrhizal types. Moreover, as a patch of vegetation changes through time, the number of associated mycorrhizal fungal species changes. For example, in one successional shrub ecosystem, each shrub expanded slowly compared with the rate of inoculum input. The mycorrhizal fungal species richness associated with any one shrub was correlated with the size of that shrub, similar to the increasing diversity of animals with increasing island size (Figure 5.3). In a large patch or where numerous plants are interacting, the variation in mycorrhizal species composition can be high. The diversity of mycorrhizal fungal species often can exceed that of the plants. Because these fungi compete with and segregate among each other, the potential resource movements can become complicated (Figure 6.6). These patterns may account for the general inability, in field studies, to predict the direction and magnitude of resource fluxes (e.g. Chiarello *et al.*, 1982; Read *et al.*, 1985).

Even 'undisturbed' ecosystems are composed of patches of disturbances within some larger matrix. Disturbance patches can be caused by a variety

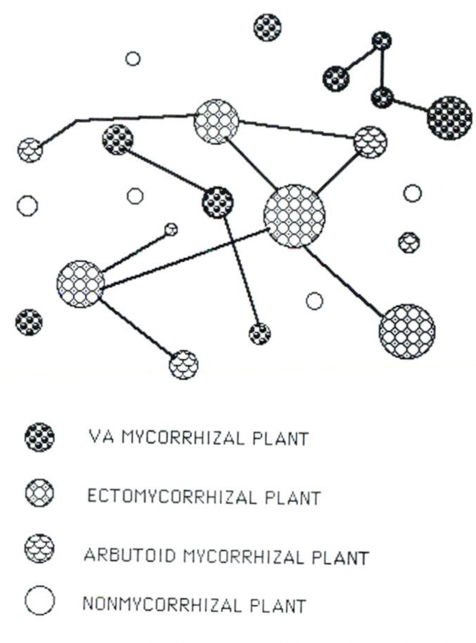

Figure 6.6. A diagram of the potential migration routes of resources in a vegetation patch with a variety of plants with differing mycorrhizal types and a variety of different possible interconnections by the hyphae.

of forces from wind to animals. They alter the mycelial matrix and the propagule density, provide sites for the establishment of exotic, immigrating inoculum, and release immobilized nutrients. Generally, disturbance patches are assumed to return the patch to an earlier successional sere. These gaps, as they are formed, usually result in a decline in mycorrhizal inoculum. Thus, patches have important consequences to the functioning of a site.

These altered conditions can have important consequences for the availability of nutrients and structure of the vegetation at a site. For example, gaps allow for the invasion and persistence of early successional species within an 'established' vegetation matrix. These often tend to be plants with different mycorrhizal characteristics from the surrounding plant species. This matrix of mycorrhizal components within an ecosystem generally results in shifts from ectomycorrhizal to VA mycorrhizal plants, from VA mycorrhizal to nonmycotrophic plants, or from ectomycorrhizal to nonmycotrophic plant patches.

A variety of animals create patches in vegetation and mycorrhizal inoculum which affect the functioning of the surrounding ecosystem. In the Bialowieza Forest of Poland, extensive soil disruptions are apparent as a result of wild boars searching for hypogeous ectomycorrhizal fungi

Figure 6.7. Ecosystem disturbances that alter the mycorrhizal inoculum density in an undisturbed site. Shown are disturbances that decrease inoculum such as (A) wild boar disturbance in the Bialowieza Forest of eastern Poland and (B) a wind throw of a burned tree in the Grand Teton National Park, Wyoming, USA, and disturbances that increase mycorrhizal inoculum including (C) gopher heaps in Nebraska (USA) grasslands

(Figure 6.7). Common invaders are VA mycorrhizal herbs or *Urtica* spp (nettles), generally considered to be a nonmycotrophic species.

Physical agents also create patches with altered inoculum conditions. Wind-throws in forest open gaps in ectomycorrhizal or VA mycorrhizal mycelial matrix thereby allowing for the entry of new plants and mycorrhizal fungi. For example, tip-up mounds from a large tree expose mineral soil without a large reservoir of mycorrhizal inoculum (Figure 6.7). These gaps exist in all forests and have unique and important features to the functioning of the forest as a whole (e.g. Pickett & White, 1985).

Gophers are known to be important regulators of ecosystems by continually turning over soils. Andersen & MacMahon (1981) estimated that gophers could completely turn over a meadow in as little as 7 years. In clearings in Douglas fir forests of the Pacific Northwest US, gopher heaps had 60 % of the spore density found in the adjacent soils (M. Allen *et al.*, 1984b). Gopher mounds in serpentine grasslands result in small mounds of low-inoculum soil available for colonization by non-mycorrhizal species (Koide & Mooney, 1987). In a coastal sage-scrub of southern California, gopher-disturbed soils contained a significantly lower diversity of mycorrhizal fungi than found in adjacent undisturbed soils (M. Allen, unpublished data). These disturbed areas contained a variety of annuals and biennials whereas the undisturbed areas were dominated by shrubs and perennial grasses.

However, these disturbances do not always reduce inoculum. In the high ash-fall areas, following the eruption of Mount St Helens, gophers brought to the surface buried soils containing mycorrhizal inoculum. These mounds of disturbed soils provided foci for the re-establishment of the facultatively mycotrophic invading plants (Figure 6.7). Also, in many instances patches survive in the midst of disturbance, forming islands of plants and mycorrhizal fungi that can colonize the surrounding disturbed matrix. On Mount St Helens, erosion zones, even in the middle of the pyroclastic flow region, were colonized rapidly by residual plants and mycorrhizal fungi within the first year after the eruption and served as sources for newly establishing vegetation (M. Allen & MacMahon, 1988). Beckjord & Hacskaylo (1984) suggested that the ectendo E-strain mycorrhizal fungus survived as a saprophyte in the rhizosphere of a VA mycorrhizal plant, *Paulownia tomentosa*, planted as buffer strips, and initiated new mycorrhizae with *Pinus strobus* following plantings. Ectomycorrhizal fungi appeared to survive harsh conditions in decayed logs following disturbance, and rapidly initiated new infections and enhanced the survival of *Tsuga heterophylla* in establishing forests (Christy *et al.*, 1982).

At some sites, there is often a mix of disturbances that increase and decrease inoculum. In a shrub-steppe ecosystem, probably 'undisturbed' since the Pleistocene, ant mounds occupied up to 20 % of the surface area (figure 6.7). The ants in these mounds clipped mycorrhizal roots from the surrounding shrubs and concentrated inoculum several orders of magnitude over the surrounding undisturbed matrix. The dominant colonizing plants after these mounds are abandoned are predominantly mycorrhizal grasses. Nearby, badgers turn over the soil in search of prey. These patches are frequently invaded by various species of Chenopodiaceae that do not form mycorrhizae.

All these perturbations provide open habitats for immigrating inoculum, whether from different plants in the undisturbed matrix immediately surrounding the disturbed patch, or from inoculum blown or carried in by an animal. Because different species of mycorrhizal fungi cause different growth responses in host plants and alter the nutrient allocation patterns among interconnected species, the importance of these patches may be considerable.

General conclusion

Mycorrhizal associations make up a large biomass component in many ecosystems. The mycorrhizae immobilize large quantities of nutrients important for decomposition and plant growth. Because mycorrhizal fungi transport much of those nutrients to the host, this immobilization is temporary and can be critical for stand production. The mycorrhizal fungal hyphae also appear to be localized where they can rapidly take up mineralized nutrients, thereby increasing the biotic regulation of nutrient cycling. Both the fungi and the associated fine roots also turn over quickly, with the result that mycorrhizae tend to be the largest throughput component, especially in forest ecosystems. No comparable data exist for other mycorrhizal systems but the mycorrhizae still remain an important constituent.

Mycorrhizae also affect the structure of ecosystems by associating with differing vegetation units (patches) within a land unit. As these patches are always changing with time, so also are the mycorrhizal density, types, and species. As no two mycorrhizae are exactly alike, the differences in these mycorrhizal patches determine many aspects of ecosystem functioning.

7

Mycorrhizae and succession

Succession is a complicated set of processes associated with vegetative recovery following disturbance. Although often set in a community ecology context, succession is more than simply a change in the plant species composition through time: it is a complex of alterations in soils, nutrient cycling, and organisms that occur following disturbance. Both types and species of mycorrhizal associations change with succession and alter processes such as organic matter development, nutrient cycling, and species composition.

Reforestation can be viewed as the priming of succession to enhance the rate of establishment of new forests. Early forestry work concentrated on the importance of mycorrhizae in forestation of regions where appropriate mycorrhizal associations did not exist. The interactions of mycorrhizae and plant succession were well recognized and hypotheses generated early in the development of mycorrhizal ecology. Thus, there may be a greater literature in the general area of mycorrhizae and succession than any other mycorrhizal ecology topic. For these reasons, I have chosen to discuss the role of mycorrhizae in a separate chapter. In this chapter, I will summarize the literature concerning mycorrhizae and succession integrating research published on dispersal, establishment, community change, and ecosystem process changes.

Mycorrhizae of different seres

Clements (1916) set the theoretical foundation for considering succession as a set of processes initiated by a disturbance and culminating in a highly organized plant community. Although best known for his 'super-organism' concept of a plant community, considered inadequate for describing plant community and population dynamics, his observations of changes in organisms and his model for organizing the processes associated with succession are useful as a tool for understanding how and why succession works (see MacMahon, 1981, for a recent discussion). Odum (1969) took a Clementsian view by proposing that entropy decreases with succession, in part because symbioses become more apparent and nutrients become associated with living or cycling biomass. These hypotheses set the stage for studying the roles of mycorrhizae in succession, an approach that currently dominates much of the mycorrhizal ecological research.

Although often perceived as recent hypotheses, the interactions of mycorrhizae and succession have deep roots. Stahl (1900) noted that many weedy plants (particularly Chenopodiaceae and Brassicaceae) were nonmycotrophic and were replaced by mycotrophic species. Dominik (1951) suggested that mycorrhizae and organic matter accumulation during succession are tightly coupled. He reported that early succesional coastal sand dunes tended to have nonmycorrhizal plants (not necessarily nonmycotrophic). As soil development proceeded and organic matter began to increase, mycorrhizal activity increased. Unfortunately, no data were presented regarding the detailed soil characteristics. Nor were roots stained so that careful observations on the successional sequence could be observed. Following his hypothesis, Frydman (1957) observed the soil development and mycorrhizal status of plants invading the ruins in Warsaw after World War II. He reported that the first invading weeds were nonmycorrhizal species, primarily Chenopodiaceae, but that VA mycorrhizae rapidly developed in the newly forming soil. In the surrounding areas with developed soils, ectomycorrhizae could be found. Nicolson (1960) observed that plants nearest the sea on coastal dunes were often nonmycotrophic (e.g. *Salsola kali*). As one proceeded inland, VA mycorrhizal infection and spore densities tended to increase and the species became more diverse (Nicolson, 1960; Nicolson & Johnston, 1979).

A set of work primarily from the coal strip mines in the United States and Great Britain in the 1960s and early 1970s suggested that the disturbance process itself tended to eliminate or severely reduce the incidence of mycorrhizal activity. Schramm (1966) noted that mycorrhizal activity was virtually absent in the slag heaps resulting from coal mining but was present in the naturally colonizing plant stands. Similar observations were made for other mining disturbances (e.g. Daft & Hacskaylo, 1976; Khan, 1978).

Observations of disturbances from other habitats and from other biomass followed these observations and linked them with earlier work on

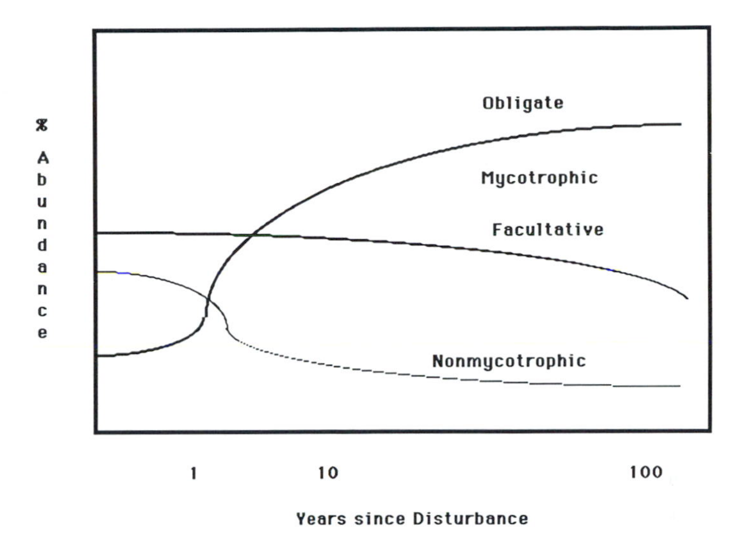

Figure 7.1. A model of succession derived from Janos (1980) using the mycorrhizal categories proposed by Stahl (1900).

successional status. Reeves *et al.* (1979) noted that the inoculum potential of human-disturbed arid lands was significantly lower than that of adjacent undisturbed areas. They further suggested that the plants invading those areas were nonmycotrophic plants, whereas mycorrhizal plants occupied the undisturbed sites. Miller (1979) reported virtually the same pattern as Reeves *et al.* (1979) in a slightly drier habitat. E. Allen & M. Allen (1980) observed that mycorrhizal activity was virtually nonexistent in recontoured spoil material left following surface mining. In sites that had topsoil replaced, mycorrhizal plant densities, mycorrhizal infection frequency, and mycorrhizal fungal spore counts rapidly recovered. However, in sites that were composed of sterile, buried materials, nonmycotrophic annual weeds predominated for up to 10 years with little or no re-establishing mycorrhizae or desirable successional grasses or shrubs. They reported that five of the seven annuals invading recontoured mine sites were nonmycorrhizal.

In a key paper, Janos (1980), working in the tropics, also stated that mycorrhizal activity may be virtually eliminated following disturbance. He further suggested that mycorrhizal activity increased during succession, associated with vegetational shifts from nonmycorrhizal plant species, to facultatively mycorrhizal plant species, to obligately mycorrhizal plant species. This model (Figure 7.1) has become the standard model for mycorrhizae and succession and truly began interest in defining the importance of mycorrhizae as regulators of interactions between successional seres.

More recent studies have modified the simple model proposed by Janos in 1980, particularly in the initial stages. Koske & Polson (1984) noted that the predominant primary dune species, *Ammophila breviligulata*, was a mycotrophic grass. They suggested that the establishment of mycorrhizae was essential to the stabilization of beach dunes and succession of the site. E. Allen & M. Allen (1990, unpublished data) found that in successional dunes on Cape Cod, the nonmycotrophic annuals were limited to colonizing the high tide lines with higher substrate nutrient concentrations (Figure 3.3). Mycorrhizal activity increased from no mycorrhizae or low colonization levels in the foredunes consisting of *Ammophila breviligulata* and beach pea (*Lathyrus japonicus*) to obligately mycorrhizal pines and oaks in the late seral dunes. Pendleton & Smith (1983) noted that, in the Great Basin of the western United States disturbed habitats had both mycotrophic and nonmycotrophic plant species. The relative dominance of the different species appeared to be associated with disturbance severity or type. The invading plants following the Mount St Helens volcanic eruption were facultative VA mycorrhizal herbs or obligately mycorrhizal trees and no nonmycotrophic species were present (M. Allen *et al.* 1984b; M. Allen, 1987a, 1988a). Schmidt & Scow (1986) reported that the flora of the Galapagos was a mix of nonmycorrhizal and VA mycorrhizal species with the nonmycotrophic plants growing in the richer, lowland soils and the VA mycorrhizal plants in the pyroclastic materials. The post-eruption species list of Isla Fernandina, Galapagos (Hendrix & Smith, 1986), also shows a mixture of nonmycotrophic and VA mycorrhizal species with many of the plants establishing as residuals in the erosion zones that presumably would have retained the VA mycorrhizal inoculum as has been found on the Mount St Helens eruption site (M. Allen *et al.*, 1984b). The first plant species to establish on the Krakatau volcano (1883) appear to be predominantly facultatively VA mycorrhizal, including grasses. However, within 20 years following the island formation, plants that approach obligate dependence on mycorrhizae (e.g. *Casuarina equisetifolia* and the ground orchid *Spathoglottis plicata*) were abundant (Simkin & Fiske, 1983). These data led E. Allen & M. Allen (1990) to hypothesize that, in different environments, mycorrhizal status and succession can vary depending on the moisture and nutrient status of the site (Figure 7.2).

Based on both of these models, following the initiation of succession, the rates and trajectories of succession should be affected by changing

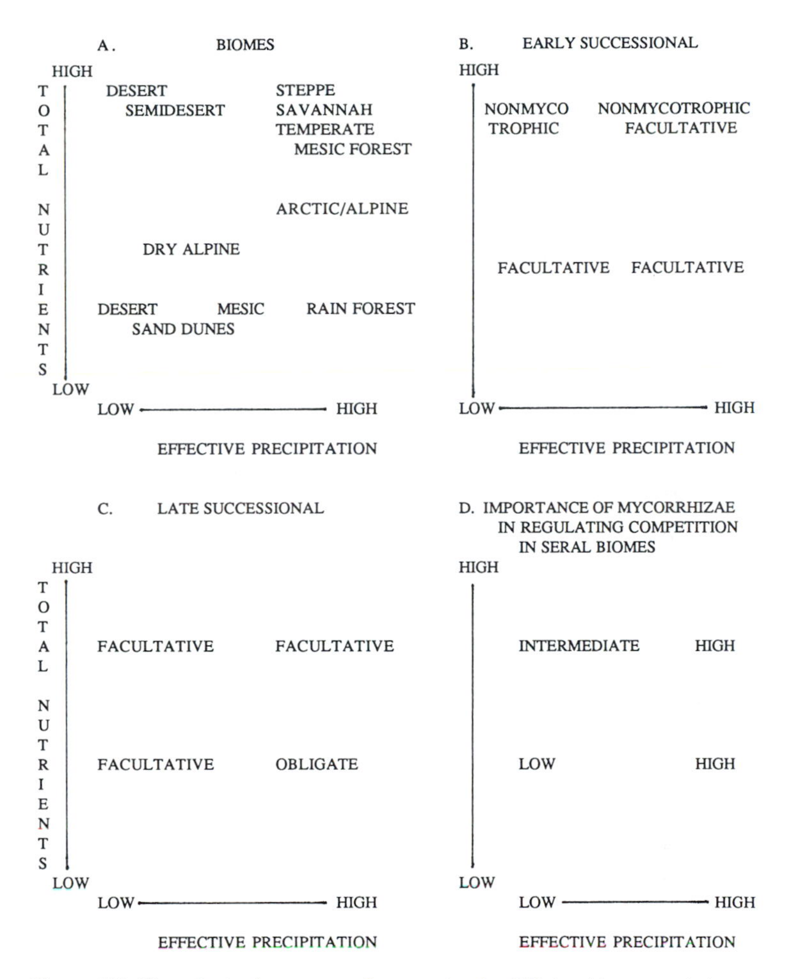

Figure 7.2. Hypothesized patterns of succession in differing biomes and the relative mycotrophy of the plants (reprinted from E. Allen & M. Allen, 1990 with permission).

mycorrhizal relationships through time. This reaction could be caused by changes in the mycorrhizal fungal symbionts, simple age structure and architecture of the organisms, and/or altered soils as a function of the successional sequence.

The roles of changing mycorrhizal fungal symbionts in affecting the successional sequence of plants have not been experimentally tested. Nevertheless, there are important species shifts with succession. Nicolson & Johnston (1979) reported that in early successional sand dunes, *Glomus fasciculatum* predominated with an increasing diversity of fungal species in later successional dunes. E. Allen *et al.* (1987) found that VA mycorrhizal fungal species diversity increased with increasing seral stage in an alpine site. Watling (1981, 1988) described a number of studies that linked changes in mycorrhizal macrofungi with changes in soils (particularly soil organic matter) and plants during long-term succession.

Several successional sequences have seres that change from VA mycorrhizal plants in the earlier seres to ectomycorrhizal plants in the later seres (e.g. West *et al.*, 1981). Schimpf *et al.* (1980) and West & Van Pelt (1987) have documented the sequence of plants with differing architectures with succession, from herbs and grasses to shrubs and trees. Each of the seres in their studies has different mycorrhizal associations.

Numerous studies have centered on changes in soil nutrients and organic matter during succession. Nitrogen, in particular, is known to be critical to succession in many habitats (e.g. Odum, 1969). All of the plants known to fix N are mycotrophic and so the association could be important to both plant survival and potentially, N accumulation. E. Allen & M. Allen (unpublished data) found that the VA mycorrhizal plants dominating the primary dunes with virtually no organic matter and low N were replaced by ectomycorrhizal and arbutoid mycorrhizal plants in the secondary and tertiary dunes with higher organic matter and N.

Other studies have documented 'successional' changes in ectomycorrhizal fungi associated with the age of a stand but with a single dominant plant species. For example, Marks & Foster (1967) noted that distinct types of fungal mantle morphotypes were replaced by other mantle types in *Pinus radiata* during a 3-year study of a 42-year-old stand. Fleming *et al.* (1984) described a sequence of mycorrhizal fungal associates through time on transplanted birch seedlings, from *Hebeloma* sp. that were initial colonizers, to *Lactarius pubescens* and *Inocybe lanuginella* that predominated during the later years of the experiment. In an experimental study of ectomycorrhizae on birch seedlings, Fleming (1985) noted that 'early-stage' colonizers such as *Hebeloma sacchariolens* and *Thelephora terrestris* persisted following transplanting whereas 'late-stage' mycorrhizal fungi such as *Amanita muscaria* and *Leccinum scabrum* did not persist and apparently required some additional soil development or late seral microbial populations in order to persist in a forest.

Mycorrhizae and patterns of successional initiation

Clearly, mycorrhizae are reduced or eliminated with disturbance and are important to the successional seres of plants invading the disturbed site. Less clear is the invasion and spreading capacity of mycorrhizal plants and fungi that subsequently affects the successional patterns described above. Only a few experimental studies have recently addressed colonization in the field and have found some complex but interesting results.

Animals have been known to feed on mycorrhizal fungi for a long time (e.g. MacMahon & Warner, 1984). Thaxter (1922) found spores of VA mycorrhizal fungi from a variety of rodents and insects. More recently, Maser *et al.* (1978b) and subsequent papers (e.g. Malajczuk *et al* ., 1987; Maser, 1988) have shown that a wide variety of rodents, in particular, transport spores from one location to another. This behavior may be especially important in the regeneration of disturbed patches in forests. Following the eruption of Mount St Helens, animals were found to move VA mycorrhizae up to several kilometers across the sterile pumice regions and initiate new mycorrhizae (M. Allen, 1988a). Both small rodents (e.g. gophers: M. Allen & MacMahon, 1988) and large ungulates (e.g. wapiti: M. Allen, 1987a) transported inoculum and initiated new infections that aided plant survival and initiated patches of new vegetation.

Even if the plant species change between seres, the fungi sometimes do not. In many regions, VA mycorrhizal plants make up the majority of plants during the entire successional chronosequence, and these mycorrhizae show little specificity. An example of this pattern includes the high plains grasslands of Wyoming in which both early reclaimed lands and undisturbed grasslands had similar species of *Glomus* present (E. Allen & M. Allen, 1980; Stahl & Christensen, 1983). Arbutoid mycorrhizal, early successional shrubs can have the same fungal associates as the later successional ectomycorrhizal trees (e.g. Perry *et al.*, 1989). The ability of these fungi to survive perturbations such as fire, as symbionts of resprouting shrubs, and to spread among plants led to the 'bootstrapping' hypothesis of Perry *et al.* (1989). This hypothesis states that disturbed ecosystems rarely re-establish 'from scratch ' (i.e. primary succession). Instead, residual soil microorganisms, especially mycorrhizal fungi, survive in patches, hold the system together, and, via their associations with several different plants, pull the system back up 'by their bootstraps'.

In many instances patches do remain in the midst of even the most devastating perturbation, forming islands of plants and mycorrhizal fungi that can colonize the surrounding disturbed matrix. On Mount St Helens, erosion zones and protected patches, even in the middle of the pyroclastic flow region, were occupied by residual plants and mycorrhizal fungi within the first year after the eruption. Surviving soil-inhabiting organisms including both gophers and ants concentrated plant and mycorrhizal

Figure 7.3. A patch of vegetation that established in the midst of the pyroclastic flow from the Mount St Helens (Washington State, USA). Plants, rodents and mycorrhizal fungi (among other organisms) survived in an erosion pocket approximately 2 km from this site and the migrating rodents initiated mycorrhizal establishment here after 3 years (M. Allen, 1988a).

fungal propagules that initiated patches of vegetation in the ash-fall zones (M. Allen *et al.* 1984b). Some of the surviving rodents even migrated across the pyroclastic flow areas and initiated new mycorrhizal symbioses with patches of initiating plants (Figure 7.3). Beckjord & Hacskaylo (1984) suggested that the 'E-strain' ectendomycorrhizal fungus survived as a saprophyte in the rhizosphere of a VA mycorrhizal plant, *Paulownia tomentosa*, planted as buffer strips. This surviving inoculum initiated new mycorrhizae with *Pinus strobus* following those plantings. Ectomycorrhizal fungi appeared to survive harsh conditions in decayed logs following disturbance, and rapidly initiated new infections and enhanced the survival of *Tsuga heterophylla* in establishing forests (Christy *et al.*, 1982). Following the establishment of a mycelial matrix between plants, nutrients and carbon can be exchanged between those interconnected plants either through direct transport from living plant to living plant via the mycelium or from dying roots of one plant to living tissue of another via the mycelium (e.g. Read *et al.*, 1985; Ritz & Newman, 1985).

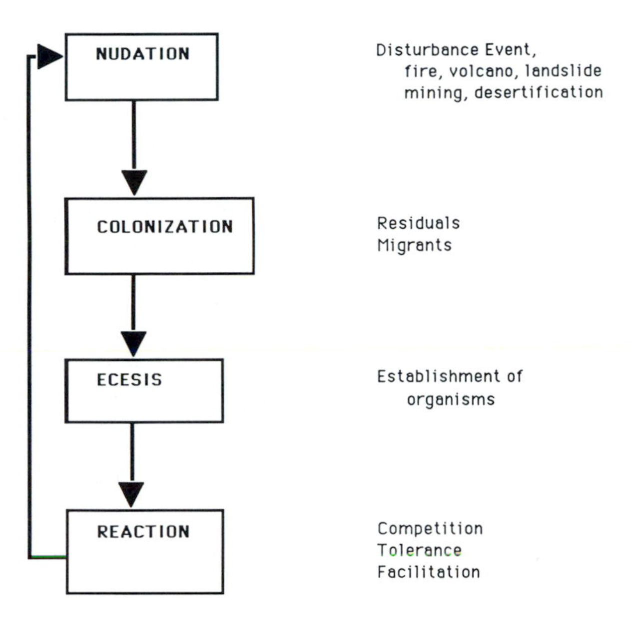

Figure 7.4. A model of succession proposed by Clements (1916) and modified by MacMahon (1981) (derived from MacMahon, 1981).

Mycorrhizae and succession in a semi-arid shrub steppe

During the years 1980–9, a major research effort to study processes regulating succession on a coal strip mine in southwestern wyoming, USA was undertaken by a group from Utah State University with the cooperation of the Kemmerer Coal Mine. The role of mycorrhizae was a major component of this effort. Since this project spawned a large number of published efforts on mycorrhizae and succession, I have devoted the remainder of this chapter to it. The site is a harsh, cold-desert sagebrush steppe at an elevation of 2200 m with summer temperatures exceeding 30 °C and winter temperatures below −30 °C . Precipitation averages *c.* 220 mm per year.

MacMahon (1981) proposed that the model of Clements (1916) regarding the steps involved in succession could be used as a set of discrete topics in which to study succession as a whole (Figure 7.4). This model was followed in the aforementioned project and those steps will form the sections that follow.

Nudation and survival of residuals

Major perturbations almost always reduce the inoculum density. The mining and soil recontouring (Fig. 7.5) at the Kemmerer Mine was an

Figure 7.5. Overview of the recontoured strip mine studied for the roles of mycorrhizae in succession. The site had been recontoured just before the photograph in 1981.

extreme case. Importantly, where old topsoil was saved, some residual inoculum survived but was redistributed randomly across the plot (Allen & MacMahon, 1985) and resulted in little to no mycorrhizal infection (E. Allen & M. Allen, 1986). The loss of spatial structure was not only limited to the ability of the fungi to colonize new plants but also led to a loss of patches with high hyphal densities and loss of spatial correlation among mycorrhizal hyphae, soil N, P, and organic matter. These changes would have major impacts on the retention of nutrients by the site as a whole (M. Allen, 1988b).

Dispersal

The primary vectors for reintroduction of mycorrhizal inoculum were hypothesized to be wind and animals. Several animals were shown to disperse inoculum onto the site (Warner *et al.*, 1987) often moving it up to 500–700 m from a source area (M. Allen 1988a). Wind also appears to be an important vector for migration of mycorrhizal fungi. Warner *et al.* (1987) demonstrated that VA mycorrhizal fungi could be wind-blown up to 2 km. M. Allen (1988a) subsequently found that animal-dispersed inoculum appeared to be far less important than wind-dispersed inoculum at this site. M. Allen *et al.* (1989b) further demonstrated that the survival of invading species of VA mycorrhizal fungi tracked closely the invasion patterns of these fungi via wind. Waaland & Allen (1987) found no apparent correlations of age of disturbance (1–30 years) and mycorrhizal activity in a cold desert. However, the native areas still had higher mycorrhizal activity and more organic matter than any other disturbed site surveyed. Mycorrhizae appeared to relate more closely to terrain and soil type.

Neither wind nor animal dispersal vectors deposited spores at random across the site. Animals deposited the majority of spores (as faeces) under or adjacent to the canopy of the shrubs (M. Allen, unpublished data). However, no apparent patterns of inoculum deposition could be found with size or location of shrubs (M. Allen, 1988a). Wind dispersal had very predictable patterns of entrainment and deposition. The planted shrubs themselves altered the turbulence patterns of winds moving at high speeds. These turbulence patterns, in turn, caused greater deposition immediately behind the shrub than in the open spaces (M. Allen, 1988b). Two shrubs, standing within 50 cm of one another, increased the deposition of spore mimics, and presumably spores, relative to isolated individuals by dramatically reducing the eddy size and vertical wind speeds and allowing the particles to drop out (L. Hipps and M. Allen, unpublished data).

These data suggest (1) that extensive dispersal of mycorrhizal fungi occurs, (2) that neither wind nor animals deposit inoculum randomly across a disturbed site, and (3) that animals and wind vectors deposited inoculum in different patterns depending on the characteristics of the establishing vegetation and edaphic factors of the site.

Ecesis and reaction: *seral stages*

The existing hypothesis is that mycorrhizae increase the competitive ability of the later successional species compared with the earlier colonizing nonmycotrophic species and thereby increase succession rates. However, in an experimental test, this hypothesis was not so straightforward (E. Allen & M. Allen, 1988; E. Allen, 1989). When plots on a high, windy ridge in a cold desert were inoculated, mycorrhizae inhibited the transition from the early seral, nonmycotrophic weeds to the later successional grasses. This appeared to be related to the environmental characteristics of the site. The predominant moisture input to this ridge is snow that, in the high winds common to the site, tends to blow away. The addition of mycorrhizae reduced the weed density and growth. The weed litter on this site served as wind-breaks trapping snow and reducing the wind damage to grass seedlings. In an analysis of five experimental units scattered across the site, two resulted in reduced succession rates with mycorrhizae, one with hastened succession rates, and two with no measurable effect on the rates (E. Allen, 1989) despite the distinct physiological benefits of mycorrhizae to the grasses and shrubs (E. Allen & M. Allen, 1986). These data indicate a need to consider the entire range of ecosystem parameters and conditions when attempting to describe the functioning of a complex symbiosis. Especially, there is a major need to examine landscape-scale phenomena such as the patterns of animal behavior during migration and wind transport of diaspores in order to understand such important factors as seed and spore sources and deposition patterns and how these affect the spatial distribution of plant communities.

Ecesis and reaction: *patch structure*

During succession mycorrhizae should not only be considered as altering competition between plants. Succession rarely proceeds in the generalized 'old-field' pattern where an entire region is considered to shift seral stages and mycorrhizal states (e.g. M. Allen, 1988a). More generally, patches of plants establish and spread followed by the establishment of new patches within the matrix of preceding vegetation. Mycorrhizal fungi, by initiating at a point of origin and spreading out between individual plants, alter the dynamics of the entire patch. When a single individual plant establishes in an open patch, the mycorrhizal hyphae radiate out from that carbon source, take up nutrients, including P, and transport those nutrients to the plant, the site of the carbon source. For example, we measured increasing patchiness of both P and VA mycorrhizal fungi with successional time. Phosphorus became more concentrated at the base of the shrub along with VA mycorrhizal spores. The absorbing hyphae were spread out from the roots into the interspace (M. Allen, unpublished data). New seedlings of both the same species and invading species then began to concentrate under and near the established, mycorrhizal shrub (M. Kelrick and J. MacMahon, unpublished data). In regions with few seral stages or with autosuccession, a limited suite of fungi can form a mycelial network that links several individuals as the vegetation increases to where plants begin to interact.

However, this site as well as most ecosystems should not be viewed as a single mycelial network. In this successional shrub ecosystem, each shrub expanded slowly compared with the rate of inoculum input. The mycorrhizal fungal species richness associated with any one shrub was

Figure 7.6. A view of the established plants on the project. The mycorrhizae have concentrated nutrients under the shrubs and the shrubs themselves have provided a focus for animal and wind dispersed invading inoculum.

correlated with the size of that shrub, similar to the increasing diversity of animals with increasing island size (M. Allen, 1988a). Specifically, in 1983, when the canopy diameter of a single shrub was only 15 cm, only *Glomus microcarpum* and *occultum* were found. By 1987, when that same shrub was 40 cm in diameter, *G. fasciculatum* and *mosseae* were also present and a nearby shrub from the undisturbed area (80 cm in diameter) had eight additional VA mycorrhizal fungal species. Small grasses and herbs growing under these canopies could be interconnected by hyphae of differing fungi in interesting and intricate patterns (see Figure 6.6).

Although additional information is still being obtained from this site, and nearby studies of other mycorrhizal researchers, it is clear that mycorrhizal associations are an important component of this ecosystem and a clear regulator of succession in both individual patches (surrounding a single shrub) and across the site as a whole (Figure 7.6).

General summary

In conclusion, mycorrhizal symbioses are major factors in the successional process. Mycorrhizal plants and fungi of the differing seral stages play off each other both directly and indirectly by altering the soil environment in which both must live. However, there remain several questions concerning how and under what conditions mycorrhizae alter the rates and direction of succession of a site. Given the wide diversity of habitats undergoing severe man-made disturbances, and the fact that natural disturbances are a natural part of the creation of new habitats and conditions, the role of mycorrhizae is still only scarcely defined.

8

Future directions for mycorrhizal research

Mycorrhizae represent one of the least understood, most widespread, and most important biological symbioses on earth. Nevertheless, much of the research is still concentrated on a relatively minor question: how can mycorrhizae overcome P deficiency in host plants? Given the wide range of responses by plants to mycorrhizae, the wide diversity of plants and fungi forming mycorrhizal associations and wide range of habitats containing mycorrhizae, the breadth of research potential and practical application of that research is almost limitless.

I would suggest that, in addition to the traditional areas of mycorrhizal research described in the previous chapters, several additional topics may broaden our conceptual understanding of the symbiosis. Interestingly, these also represent the two ends of the biological hierarchy. They are the molecular biology of mycorrhizae and the relationships between mycorrhizal fungus, plant and environment at the landscape to global scale.

What determines a mycorrhiza?

One of the difficulties in working with mycorrhizae is determining exactly what a mycorrhiza is. The definition of a mycorrhiza is functional, i.e. a mutualistic association between plant and fungus, but the identification of a mycorrhiza is generally structural, e.g. the presence of a hartig net, arbuscules, vesicles, etc. Much of this contradiction lies in the inexact distinctions among types of symbioses and the inability to identify mechanisms whereby two interacting organisms coevolve to enhance the fitness of both. Harley (1968) suggested that there is really a gradient in interactions between plant and fungus from parasitic to mutualistic. And, since mycorrhizal symbioses are not generally species-specific (which characterizes much of the theoretical thinking regarding coevolution), the fungus often acts as a parasite in time (e.g. the seedling stage) or in space (degradation of specific roots or parts such as root hairs). Generally, in these cases, the presence of the mycorrhiza is thought to increase the fitness of the host plant and therefore should be considered mutualistic.

However, few mycorrhizal associations can be studied for the length of time and range of conditions that allow us to denote when the association is mutualistic (and therefore mycorrhizal) and when the association is parasitic (and therefore nonmycorrhizal). If Harley's (1968) continuum from mutualism to parasitism is true, then determining a mycorrhiza simply by looking for specific structures (e.g. Hirrel *et al.*, 1978) is not valid. For example, in the case of *Salsola kali*, the mycorrhizal fungus invades the roots of a seedling where it forms all appropriate structures. However, the plant does not gain either physiological or growth benefits and seedling mortality may result (M. Allen *et al.*, 1989a). Alternatively, some species of Brassicaceae including *Brassica napus* and *Brassica nigra* were unaffected by the presence of the fungus as the hyphae appear to bypass the plant entirely (Glenn *et al.*, 1988; B. Weinbaum *et al.*, unpublished observations), and in another experiment, *Brassica napus* had a different response in that the fungus reportedly colonized the plant but at infrequent levels (Tommerup, 1984; Glenn *et al.*, 1985). This is important beyond just the implications for the effects of mycorrhizal fungi on single plants. For example, mycorrhizae appear to regulate succession in at least one shrub-steppe ecosystem, not by dramatically improving growth of mycotrophic plants, but by reducing the fitness of a competing nonmycotrophic plant (E. Allen & M. Allen, 1988). A functional rather than a structural approach is needed to understand whether a fungus–plant association is a mycorrhiza (mutualist) or a parasite.

A functional approach may be taken by using molecular techniques to identify some of the genetic basis for mycorrhizal associations. Plant–microbe molecular biologists have looked for the molecular signals whereby plants and microbes recognize each other. Particularly, how do plants recognize an invading pathogen and protect themselves against that parasite, and how does a microorganism recognize a host that it can successfully invade, establish on and successfully reproduce in? They suggest that these signals have a genetic basis consistent with co-evolutionary theory.

Anderson (1988) reviewed these concepts and suggested how they might apply to mycorrhizae. She suggested that an elicitor model might apply to mycorrhizal mutualisms as well as plant parasitisms. Specifically, this model states that any mycorrhizal fungus will invade a nearby plant and, if the plant fails to recognize it, the mycorrhiza will form. Alternatively, if the fungus is recognized, the plant will attempt to reject it as a parasite. A. Anderson (personal communication) recently found elicitors in bean (*Phaseolus vulgaris*) (a VA mycorrhizal plant) for ectomycorrhizal fungi, an indication that mycorrhizal plants react to an incompatible mycorrhizal fungus. M. Allen *et al.* (1989a) found that *Salsola kali*, a nonmycotrophic weed, reacted to challenges by VA mycorrhizal fungi but that *Agropyron smithii*, a mycotrophic grass, did not. These limited data suggest that a mycorrhiza could be defined, in part, by the failure of a host to reject a known mycorrhizal fungus (M. Allen *et al.*, 1989a).

The molecular biologists have developed powerful genetic theories and techniques that may be able to aid in our understanding of mycorrhizal symbioses. The ability to understand the recognition processes between fungus and plant will aid our ability to describe the coevolutionary interactions that shape the nature of mycorrhizal symbioses. Too often, in mycorrhizal research, the search for increased plant growth or 'improved' physiological traits have been the object in our search for mechanisms of mycorrhizal action. If we accept the premise that mycorrhizae are coevolved fungi and plants, then looking for the genetic characteristics of mycorrhizae such as recognition, compatibility, and changes through time, may be the most important new direction for mycorrhizal population ecology.

Tracking mycorrhizal fungi

As discussed in Chapter 5 in detail, there is no single 'mycorrhizal' response as there are numerous species of mycorrhizal fungi representing several different taxa. However, because of the inability to distinguish the differing fungal taxa (especially as hyphae), even knowing if a particular fungus is present is extremely difficult. Describing the effects of the differing fungi over the life span of a plant or even a short-term experiment in the field is almost impossible. Several attempts have been made to create keys to the differing morphotypes of ectomycorrhizal fungi based on the morphology of the root–fungus interface (e.g. Wilcox, 1982). Similar efforts have been made with VA mycorrhizal fungi (e.g. Abbott & Robson, 1979). Also, extensive studies have attempted to inoculate a plant with one fungus, plant it into the field, and ascribe the resulting responses to that fungus.

The use of electrophoresis for differentiating fungi indicates that specific isozymes can be used to distinguish both species and populations (Hepper *et al.*, 1986). They can be used to differentiate both spores and the fungal hyphae within roots (Hepper *et al.*, 1988) as well as to track the relative interactions among mycorrhizal fungi within those roots (Hepper *et al.*, 1988). However, these procedures are time-consuming and can only be used for small root segments and limited numbers of fungi.

Several groups have recently begun to develop techniques that would allow for the rapid assessment of specific fungi in the vegetative state in association with the host plant. Such techniques as application of monoclonal and polyclonal antibodies to the fungal walls offer techniques for rapid assessment of specific fungi that invade roots (Figure 8.1). However, often the specificity is limited, particularly in the case of the polyclonal antibodies (Kough *et al.*, 1983; Wilson *et al.*, 1983). Nevertheless, this method is inexpensive and the technique can be widely used (Friese & M. Allen, 1990). Monoclonal antibodies, while much more specific (Wright *et al.*, 1987), are expensive to develop. Nevertheless, they are specific enough (purportedly to the species level) to reduce the limited value of the polyclonal antibody procedure.

Recently, several groups of researchers have begun to look for restriction fragment length polymorphisms (RFLPs). These appear to be highly specific for given populations and have widespread possibilities for the future.

All of these efforts and more are desperately needed before an understanding of the ecology of the differing mycorrhizal fungal species and ecotypes, and their effects on their surrounding environment, can be substantially furthered. Greater interactions with molecular biologists and their powerful techniques are critical for furthering our understanding and application of mycorrhizal technology.

Mycorrhizae and the landscape

The recent emergence of landscape ecology, if not as a separate discipline, then certainly as an important perspective in ecology, has lead to important ideas of integrating the dynamics of interactive habitats. Several important papers suggested that the dynamics across landscapes affect ecosystem processes. These include the interspersion of communities across a land unit (Forman & Godron, 1986), the importance of boundaries and flows of materials and propagules (Weins *et al.*, 1986), and the differing effects of mycorrhizae on succession across a complex terrain (E. Allen, 1989). All indicate the mycorrhizal action should be placed in a larger spatial framework to describe their importance in both native and managed lands. The importance of these becomes apparent as one looks across almost any landscape (Figure 8.2). Three areas of study of mycorrhizal-regulated processes have especially dramatized the import-

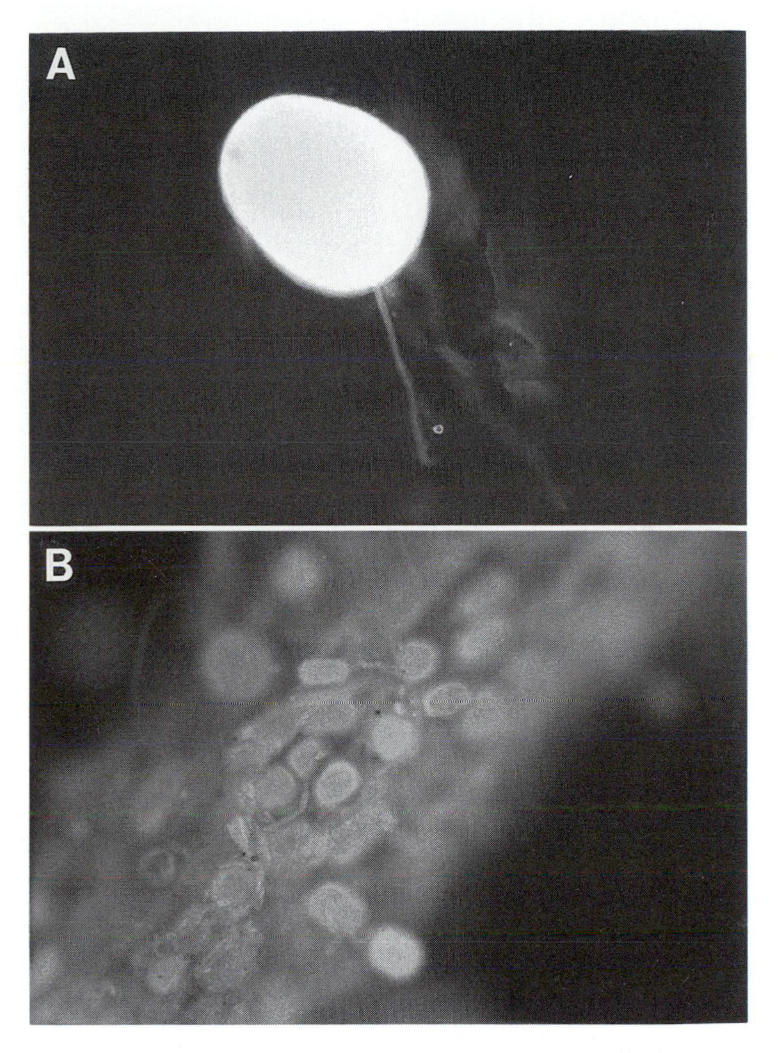

Figure 8.1. Comparative response of *Gigaspora margarita* (A) and *Glomus etunicatus* (B) to fluorescing antibodies produced to *Gigaspora margarita*. The antibodies appear at least to be specific to the genus level and can distinguish vegetative hyphae as well as spores (Friese & Allen, 1990).

ance of this perspective. These include (1) the debate over the 'mechanism of action' of VA mycorrhizae, (2) the restoration of mycorrhizae in disturbed lands, and (3) the effects of atmospheric pollution on mycorrhizae and vegetation.

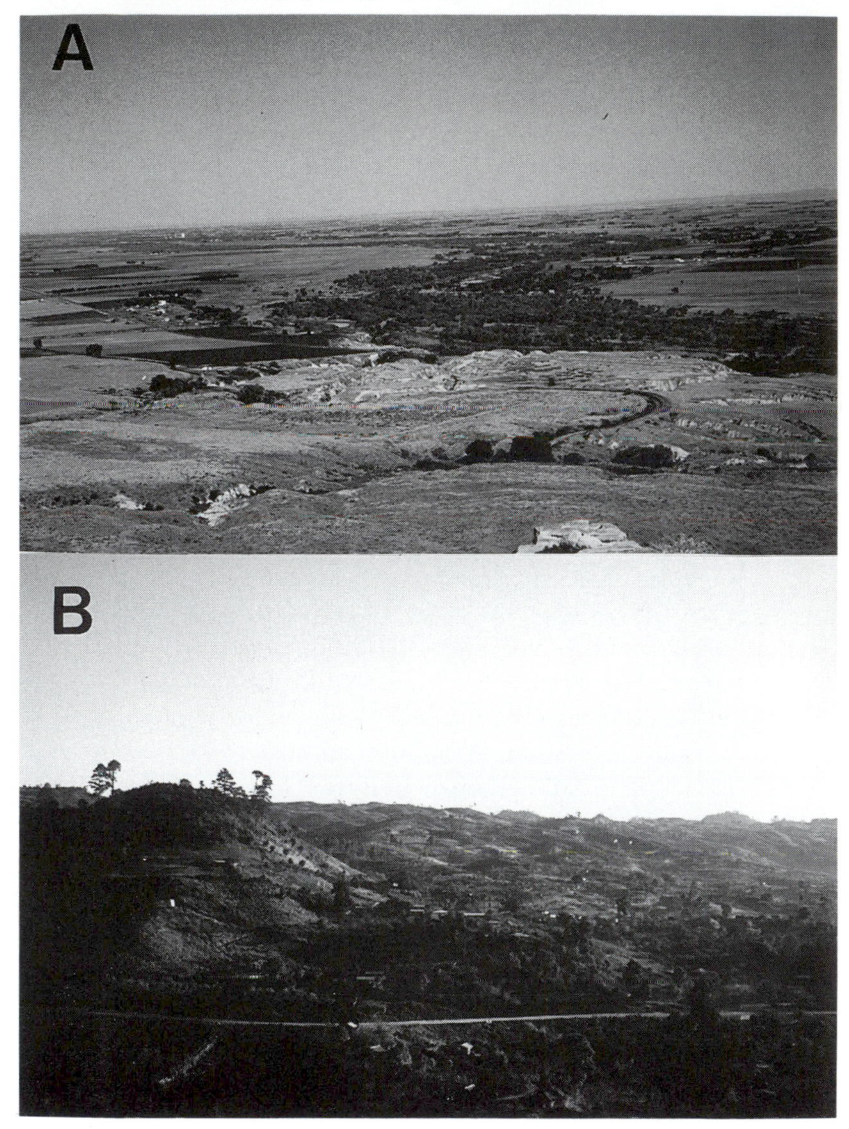

Figure 8.2 Two landscape overviews. Photograph A is an overview of the Platte Valley in western Nebraska, USA. Included in this scenery is the native grassland (VA mycorrhizal plants bisected by the riparian habitat of the Platte River which is dominated by ectomycorrhizal *Populus deltoides* and *Salix* spp.) Superimposed on this landscape are urban neighborhoods with ectomycorrhizal trees (e.g. *Ulmus*, *Pinus*, *Picea*) and a variety of horticultural plants with differing mycorrhizal associations, VA mycorrhizal wheat (*Triticum aestivum*) fields, and nonmycorrhizal sugar beet (*Beta vulgaris*) fields. Photograph B shows an overview of an ectomycorrhizal *Pinus* forest that is being cleared to plant VA mycorrhizal *Zea mays* in the Mexico central highlands.

Mechanism of VA mycorrhizal action

Discussions over the means whereby VA mycorrhizae affect plants have arisen from two viewpoints. In the first, derived from production agriculture studies, workers have hypothesized that VA mycorrhizae increase P uptake by increasing absorbing surface area and that all other responses are secondary (e.g. Sanders & Tinker, 1973; Graham, 1987; Fitter, 1988). In this context there are calls for standardizing the growth medium and plant species to describe this mechanism further. If this viewpoint is valid, it would have implications at the scale of the landscape as described below.

The second viewpoint arose primarily from restoration ecology which has encouraged the study of VA mycorrhizal responses in a variety of habitats to describe the need for maintaining or re-introducing VA mycorrhizae to sites following severe land disturbance or degradation. Specifically, several researchers, working in different habitats, have suggested that varied responses of plants to mycorrhizae can characterize the mutualism depending on the environmental conditions. For example, VA mycorrhizae may improve water uptake in arid to semi-arid habitats that generally contain high levels of soil P (e.g. Levy & Krikun, 1979; M. Allen *et al.*, 1981b; M. Allen, 1982; Bildusas *et al.*, 1986; Trent *et al.*, 1989), hormone balance in response to stress (M. Allen 1980, 1982) and direct N transport when soils contain predominantly NH_4^+-N (Ames *et al.*, 1983). When this variety of habitats was studied, many contained high quantities of soil P, in both available and unavailable forms. N. Warner and M. Allen (unpublished data) even reported high VA mycorrhizal infection of plants growing on mined phosphate being prepared for milling with up to 200 mg per kg P.

Both viewpoints become especially valuable in the context of theory and land management. If the action of VA mycorrhizae could be reduced to a single factor, such as improved P nutrition, then either additional mycorrhizae or P fertilizers could be applied to improve plant production where needed but not added where soils are not P deficient. In areas with high P, a reduction in VA mycorrhizal activity could be valuable as in soils with high P, VA mycorrhizae can reduce seedling growth (Buwalda & Goh, 1982; Bethlenfalvay *et al.*, 1982). If P were the only VA mycorrhizal benefit, the added energy drain of maintaining a mycorrhiza would indicate that in soils of adequate or high P, mycorrhizae should be eliminated to increase production and should be considered not as mutualisms but rather as parasites with a limited mutualistic capacity. The fact that natural selection does not eliminate VA mycorrhizae or eliminate many of the benefits of the symbiosis to the host plants in areas of high P argues against this viewpoint.

If the action of VA mycorrhizae were variable depending on resource limits, the association would be a mutualism for many if not most cases. Mycorrhizae could improve productivity or fitness in soils with both high and low P. In this scenario, one might expect to find ecotypic differentiation of fungi from different sites. These differences could be used for reclaiming a single site with a complex of environmental conditions or could be used differentially to reclaim differing sites. This condition would also necessitate a broad range of studies designed to differentiate the variable and multiple effects of all pertinent ecosystem processes and properties as one moves across boundaries created by soils, species, topography, or other factors.

While the latter scenario is much more complicated, it seems trite to assume that all interactions between members of two distinct kingdoms could be ascribed to one simple element. A much more valuable and realistic approach is to look for the mechanisms whereby the multitude of mycorrhizal fungi and plants interact in differing habitats. These may or may not be of economic value but would surely increase our basic understanding of mycorrhizal symbioses.

Restoration of mycorrhizae on disturbed lands

Restoration of disturbed lands has prompted a need for understanding mycorrhizae at the landscape scale. Early studies on mycorrhizae and lands disturbed for mineral extraction found that severe land disturbance reduced or eliminated mycorrhizal fungi and that many of the desirable plant species were dependent on mycorrhizae to establish and survive (e.g. Schramm, 1966; Marx, 1975; Reeves *et al.*, 1979; E. Allen & M. Allen, 1980). Into the 1980s, only a limited suite of ectomycorrhizal fungi were presumed to be wind-dispersed. Most ectomycorrhizal fungi and VA mycorrhizal fungi had been demonstrated to migrate via animal vectors (e.g. Marx, 1975; Maser *et al.*, 1978a; MacMahon & Warner, 1984). Thus recovery of mycorrhizae was presumed to be limited to the extension of animal activity as plants and animals slowly recolonized the site from the edges of the disturbance. This led to research efforts aimed at inoculating plants in disturbed areas (e.g. Marx, 1975; Wood, 1984).

More recently, mycorrhizal fungi were demonstrated to move rapidly and widely across disturbed areas by both wind and animals (M. Allen,

1987b, 1988a; Warner *et al.*, 1987). Animals and wind both respond to topographic barriers. The pattern of dispersal of the fungi and their host plants is a function of the responses of the vector to the topographic features and the extent to which the vector migrates. M. Allen (1988a) described the patterns of VA mycorrhizal fungal spore migration onto two sites, a windy shrub-steppe and the false alpine of the Mount St Helens eruption zones, using the migration patterns described by Forman & Godron (1986). At the windy site, spores were predominantly dispersed by wind and deposited across the site depending on the wind dynamics around the complex terrain (M. Allen *et al.*, 1989b). Interestingly, the fungi from the two distinct source areas had different depositional patterns (M. Allen *et al.*, 1989b) and also had distinctly different effects on a host plant when grown in a common garden experiment (Hinckley & Christensen, 1989). On Mount St Helens, animals dispersed the spores from vegetation patch to vegetation patch as plants reestablished across the site (M. Allen, 1988a).

A combination of describing the dispersal vectors for a given habitat, knowing the migration patterns of those vectors, and knowing how those vectors deposit the spores should allow for the prediction of the patterns of mycorrhizal immigration. Moreover, by knowing these patterns and the characteristics of the fungi from the source regions, one should be able to predict the types of fungi that will immigrate. Application of this information also could create the potential for altering the species compositions. By artificially inoculating a given site where a vector would move diaspores, one could initiate the reestablishment of a desirable mycorrhizal fungus across a given landscape. Experiments to test these hypotheses have not been completed and will be important in describing the successional patterns on a given site as well as improving the revegetation rates. Coupling land management and the ability to screen or engineer mycorrhizal fungi with specific desirable characters could improve the range of tools available to land managers.

Mycorrhizae and air pollution

Any factor that alters the soil environment physically or chemically is bound to affect the mycorrhizal fungi. A clear case of an important result of our industrial age is the forest die-back which covers extensive areas in Europe and is beginning to affect North American forests (Figure 8.3). This forest decline is associated with acid precipitation input although the specific mechanisms involved are still highly debated.

Tracking and describing the effects of air pollution, from specific

Figure 8.3. Forest decline in the Black Forest of West Germany. Among the proposed reasons for this decline is the reduction in diversity and density of mycorrhizal fungi (see text).

substances such as ozone to unbounded, reactive products such as acid rain depends on understanding the pathways of the air-dispersed components and their reactions with the biota. Pollution from industrial, automotive, and agricultural sources affects many life forms and often in indirect ways that are difficult to quantify. Mycorrhizal fungi may play an important role in understanding the causes and, potentially, the restriction, of air pollution damage. However, delineating the functioning of mycorrhizae depends on understanding the importance and responses of mycorrhizae to pollution over broad land expanses.

One of the best documented effects of industrial pollution on forest health is from the Polish forests. Kowalski (1987) noted that in areas of strong industrial pollution, the activity and diversity of mycorrhizal fungi was dramatically reduced compared with areas with lesser pollution. In areas of high pollution, structurally 'primitive' mycorrhizae were observed that often had parasitic characteristics. Outplantings of seedlings with well developed mycorrhizae were more resistant to pollution than trees with poorly developed or no mycorrhizae. These data strongly suggest a 'Catch-22' relationship between pollution and mycorrhizae. It takes well developed mycorrhizae to tolerate pollution but mycorrhizal activity and diversity are reduced with pollution.

These observations relate closely to the observations of other regions of the world. The mountains of central Britain that had been forested prior to the Industrial Revolution never recovered, in part because of the heavy metal pollution resulting from that era. Many of those areas are currently dominated by heath plants whose ericoid mycorrhizae are highly tolerant of heavy metals (e.g. Bradley *et al.*, 1981, 1982).

Mycorrhizae and global ecology

Understanding the linkages between biosphere and the atmosphere may be the most challenging problem of the 1990s for ecologists. Possibly the single most critical problem is that of potential global climate warming in response to increasing atmospheric CO_2. As discussed in earlier chapters, mycorrhizae are responsible for processing a large fraction of the carbon cycled in terrestrial ecosystems. This processing ranges from regulating the total amount of carbon fixed, either during photosynthesis or via direct dark fixation by the fungi, to processing as much as 40 % of the carbon respired and immobilized in the infamous 'below-ground biomass'. In addition, mycorrhizal associations appear to have evolved during periods of higher CO_2 than is present today. These facts imply that there is a large and valuable role for understanding mycorrhizal dynamics in response to climate change.

There are several ways in which the roles of mycorrhizae could be studied. These include assessing the importance of mycorrhizae in carbon cycling, the impacts of CO_2 fertilization on mycorrhizae and mycorrhizal-nutrient cycling, and the potential changes in mycorrhizal relationships in response to temperature and moisture changes.

The first, assessing the contribution of mycorrhizal fungi, has not yet been adequately attempted. Values have varied from 1 % of the photosynthetically fixed CO_2 (Harris & Paul, 1987) to well over 30 % of

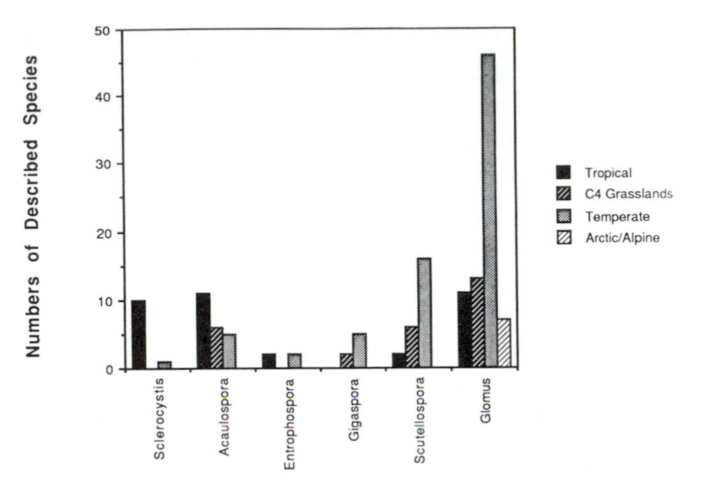

Figure 8.4. Locations where the species of the genera of VA mycorrhizal fungi were described (taken from the descriptions reprinted in Schenck and Perez, 1988). Locations were divided into tropics, C_4 grasslands (derived from the immigration of tropical grasses into temperate regions) plus subtropical areas, and temperate regions. The resulting plot suggests that these fungi were derived in differing habitats. Their abilities to tolerate major changes in climates remain unknown.

the net primary production (St John & Coleman, 1983). These estimates badly need to be refined if the solution of the second problem, the effects of CO_2 fertilization, is to be attempted. What few data exist suggest that the mycorrhizae could respond rather dramatically to CO_2 fertilization. O'Neill & Norby (1988) noted that elevated CO_2 increased ectomycorrhizal infection percentages and O'Neill *et al.* (1989) reported that enriched CO_2 also increased rooting density while maintaining an equal percentage of VA mycorrhizal infection, which means an increase in total mycorrhizae per plant and per unit soil volume. If increased productivity should result, nutrients and water, soil resources provided by mycorrhizae, could become limiting more rapidly as they are depleted and the need for more and more efficient mycorrhizal fungi could increase.

Finally, mycorrhizal plants and fungi could be dramatically affected by changing world climate as would all other organisms. Mycorrhizal fungi, as other organisms, have evolved and adapted to specific habitats. Studies on the biogeography of ectomycorrhizal fungi have been integrated well into surveys on the distribution of these fungi (e.g. Horak, 1983). However, few similar attempts have been made for VAs (for exceptions, see the work of Koske and colleagues, e.g. Koske, 1987). Nevertheless, even these fungi have important dispersion patterns. For example, simply surveying the locations where species have been described reveals some interesting patterns (Figure 8.4). Many fungi probably have restricted habitats just as do many plants. As climates change, the ability of the mycorrhizal fungi to change or disperse as well as the responses of the host plants could be a major determinant in the ability to maintain productive ecosystems.

Conclusion

In 1885, Professor Frank was commissioned by the King of Prussia to determine how to grow truffles. He failed. It was almost a century before the discoveries that he made were realized as the goals of his original project. What he did was to describe and initiate research on a small fragment of a root with a fungal association, a mykorhiza, that was to revolutionize our understanding of the importance of symbiosis and below-ground organisms in virtually all terrestrial ecological processes. Theoretical research on mycorrhizae, ranging from the molecular clues that fungus and plant use to distinguish a partner from an antagonist to why ecosystems conserve or lose soil resources, is expanding at an exponential rate. Research on mycorrhizae in reclaimed mined lands led, in part, to many revised laws requiring careful management of topsoil and continuing efforts by government agencies to restore disturbed habitats.

We all should fail so nobly.

References

Abbott, L. K. & Robson, A. D. (1979). A quantitative study of the spores and anatomy of mycorrhizas formed by a species of *Glomus*, with reference to its taxonomy. *Australian Journal of Botany*, **27**, 363–75.

Abbott, L. K. & Robson, A. D. (1984). Colonization of the root system of subterranean clover by three species of vesicular-arbuscular mycorrhizal fungi. *New Phytologist*, **96**, 275–81.

Abbott, L. K. & Robson, A. D. (1985). Formation of external hyphae in soil by four species of vesicular-arbuscular mycorrhizal fungi. *New Phytologist*, **99**, 245–55.

Abuzinadah, R. A., Finlay, R. D. & Read, D. J. (1986). The role of proteins in the nitrogen nutrition of ectomycorrhizal plants. II. Utilization of protein by mycorrhizal plants of *Pinus contorta*. *New Phytologist*, **103**, 495–506.

Abuzinadah, R. A. & Read, D. J. (1986). The role of proteins in the nitrogen nutrition of ectomycorrhizal plants. III. Protein utilization by *Betula*, *Picea* and *Pinus* in mycorrhizal association with *Heboloma crustiliniforme*. *New Phytologist*, **103**, 507–14.

Aldon, E. F. (1975). Endomycorrhizae enhance survival and growth of fourwing saltbush on coalmine spoils. *USDA Forest Service Research Note R.M.* 294.

Alexander, C., Alexander, I. J. & Hadley, G. (1984). Phosphate uptake in relation to mycorrhizal infection. *New Phytologist*, **97**, 401–11.

Alexander, C. & Hadley, G. (1984). The effect of mycorrhizal infection of *Goodyera repens* and its control by fungicide. *New Phytologist*, **97**, 391–400.

Alexander, C. & Hadley, G. (1985). Carbon movement between host and mycorrhizal endophyte during the development of the orchid *Goodyera repens* Br. *New Phytologist*, **101**, 657–65.

Alexander, I. J. & Hardie, K. (1981). Surface phosphatase activity of Sitka spruce mycorrhizas from serpentine sites. *Soil Biology and Biochemistry*, **13**, 301–5.

Alexander, T. & Weber, H. C. (1985). The parasitic way of life of *Parentucellia latifolia*, Scrophulariaceae. *Bietraege zur Biologie der Pflanzen*, **60**, 23–34.

Alexopoulos, C. J. & Mims, C. W. (1979). Introductory Mycology, *3rd edition*. John Wiley & Sons, New York.

Allee, W. C., Emerson, A. E., Park, O., Park, T. & Schmidt, K. P. (1949). *Principles of Animal Ecology*. W. B. Sanders Company, Philadelphia.

Allen, E. B. (1984). The role of mycorrhizae in mined land diversity. In *Proceedings of the 3rd biennial symposium on surface coal mine reclamation of the Great Plains*, Billings, Montana, pp. 273–95. Montana State University, Bozeman, MT.

Allen, E. B. (1989). The restoration of disturbed arid landscapes with special reference to mycorrhizal fungi. *Journal of Arid Environments*, **17**, 279–86.

Allen, E. B. & Allen, M. F. (1980). Natural re-establishment of vesicular-arbuscular mycorrhizae following stripmine reclamation in Wyoming. *Journal of Applied Ecology*, **17**, 139–47.

Allen, E. B. & Allen, M. F. (1984). Competition between plants of different successional stages: mycorrhizae as regulators. *Canadian Journal of Botany*, **62**, 2625–9.

Allen, E. B. & Allen, M. F. (1986). Water relations of xeric grasses in the field: interactions of mycorrhizae and competition. *New Phytologist*, **104**, 559–71.

Allen, E. B. & Allen, M. F. (1988). Facilitation of succession by the nonmycotrophic colonizer *Salsola kali* (Chenopodiaceae) on a harsh site: effects on mycorrhizal fungi. *American Journal of Botany*, **75**, 257–66.

Allen, E. B. & Allen, M. F. (1990). The mediation of competition by mycorrhizae in successional and patchy environments. In *Perspectives on Plant Competition*, ed. J. B. Grace and G. D. Tilman, pp. 367–89. Academic Press, New York.

Allen, E. B., Chambers, J. C., Conner, K. F., Allen, M. F. & Brown, R. W. (1987). Natural reestablishment of mycorrhizae in disturbed alpine ecosystems. *Arctic and Alpine Research*, **19**, 11–20.

Allen, E. B. & Cunningham, G. L. (1983). Effects of vesicular-arbuscular mycorrhizae on *Distichlis spicata* under three salinity levels. *New Phytologist*, **93**, 227–36.

Allen, M. F. (1980). Physiological alternations associated with vesicular-arbuscular mycorrhizal infection in *Bouteloua gracilis*. PhD dissertation, University of Wyoming.

Allen, M. F. (1982). Influence of vesicular-arbuscular mycorrhizae on water movement through *Bouteloua gracilis* (H.B.K.) Lag ex Steud. *New Phytologist*, **91**, 191–6.

Allen, M. F. (1983). Formation of vesicular-arbuscular mycorrhizae in *Atriplex gardneri* (Chenopodiaceae): seasonal response in a cold desert. *Mycologia*, **75**, 773–6.

Allen, M. F. (1985). Phytohormone action: an integrative approach to understanding diverse mycorrhizal responses. In *Proceedings of the 6th North American Conference on Mycorrhizae*, ed. R. Molina, pp. 158–60. Forest Sciences Laboratory, Corvallis, OR.

Allen, M. F. (1987a). Re-establishment of mycorrhizas on Mount St Helens: migration vectors. *Transactions of the British Mycological Society*, **88**, 413–17.

Allen, M. F. (1987b). Ecology of vesicular-arbuscular mycorrhizae in an arid ecosystem: use of natural processes promoting dispersal and establishment. In *Mycorrhizae in the Next Decade, Practical Applications and Research Priorities*, ed. D. M. Sylvia, L. L. Hung and J. H. Graham, pp. 133–5. IFAS, Gainesville, FL.

Allen, M. F. (1988a). Re-establishment VA of mycorrhizae following severe disturbance: comparative patch dynamics of a shrub desert and a subalpine volcano. *Proceedings of the Royal Society of Edinburgh*, **94B**, 63–71.

Allen, M. F. (1988b). Belowground spatial patterning: influence of root architecture, microorganisms and nutrients on plant survival in arid lands. In *The Reconstruction of Disturbed Arid Lands: an Ecological Approach*, ed. E. B. Allen, pp. 113–135. Westview Press, Boulder, CO.

Allen, M. F. & Allen, E. B. (1990). Carbon source of VA mycorrhizal fungi associated with Chenopodiaceae from a semi-arid steppe. *Ecology*, **71**, 2019–21.

Allen, M. F., Allen, E. B. & Friese, C. F. (1989a). Responses of the non-mycotrophic plant *Salsola kali* to invasion by vesicular-arbuscular mycorrhizal fungi. *New Phytologist*, **111**, 45–9.

Allen, M. F., Allen, E. B. & Stahl, P. D. (1984a). Differential niche response of *Bouteloua gracilis* and *Pascopyrum smithii*: to VA mycorrhizae. *Bulletin of the Torrey Botanical Club*, **111**, 316–25.

Allen, M. F., Allen, E. B. & West, N. E. (1987). Influence of parasitic and mutualistic fungi on *Artemisia tridentata* during high precipitation years. *Bulletin of the Torrey Botanical Club*, **114**, 272–9.

Allen, M. F. & Boosalis, M. G. (1983). Effects of two species of VA mycorrhizal fungi on drought tolerance of winter wheat. *New Phytologist*, **93**, 67–76.

Allen, M. F., Hipps, L. E. & Wooldridge, G. L. (1989b). Wind dispersal and subsequent establishment of VA mycorrhizal fungi across a successional arid landscape. *Landscape Ecology*, **2**, 165–7.

Allen, M. F. & MacMahon, J. A. (1985). Importance of disturbance on cold desert fungi: comparative microscale dispersion patterns. *Pedobiologia*, **28**, 215–24.

Allen, M. F. & MacMahon, J. A. (1988). Direct VA mycorrhizal inoculation of colonizing plants by pocket gophers (*Thomomys talpoides*) on Mount St. Helens. *Mycologia*, **80**, 754–6.

Allen, M. F., MacMahon, J. A. & Andersen, D. C. (1984b). Reestablishment of Endogonaceae on Mount St Helens: survival of residuals. *Mycologia*, **76**, 1031–8.

Allen, M. F., Moore, T. S., Jr & Christensen, M. (1980). Phytohormone changes in *Bouteloua gracilis* infected by vesicular mycorrhizae. I. Cytokinin increases in the host plant. *Canadian Journal of Botany*, **58**, 371–74.

Allen, M. F., Moore, T. S., Jr & Christensen, M. (1982). Phytohormone changes in *Bouteloua gracilis* infected by vesicular mycorrhizae. II. Altered levels of gibberellin-like substances and abscisic acid in the host plant. *Canadian Journal of Botany*, **60**, 468–71.

Allen, M. F., Richards, J. H. & Busso, C. A. (1989c). Influence of clipping and soil water status on vesicular-arbuscular mycorrhizae of two semiarid tussock grasses. *Biology and Fertility of Soils*, **8**, 285–9.

Allen, M. F., Sexton, J. C., Moore, T. S., Jr & Christensen, M. (1981a). Influence of phosphate source on vesicular-arbuscular mycorrhizae of *Bouteloua gracilis. New Phytologist*, **87**, 687–94.

Allen, M. F., Smith, W. K., Moore, T. S., Jr & Christensen, M. (1981b). Comparative water relations and photosynthesis of mycorrhizal and non-mycorrhizal *Bouteloua gracilis* (J.B.K.) Lag ex Steud. *New Phytologist*, **88**, 683–93.

Amaranthus, M. P. & Perry, D. A. (1989). Interaction between vegetation type and madrone soil inocula in the growth, survival, and mycorrhizal formation of Douglas fir. *Canadian Journal of Forest Research* (in press).

Ames, R. N. (1987). Mycorrhizosphere morphology and microbiology. In *Mycorrhizae in the Next Decade, Practical Applications and Research Priorities*, ed. D. M. Sylvia, L. L. Hung and J. H. Graham, pp. 181–3. IFAS, Gainesville, FL.

Ames, R. N., Reid, C. P. P., Porter, L. K. & Cambardella, C. (1983). Hyphal uptake and transport of nitrogen from two ^{15}N-labelled sources by *Glomus mosseae*, a vesicular-arbuscular mycorrhizal fungus. *New Phytologist*, **95**, 381–96.

Anagnostakis, S. L. (1987). Chestnut blight: the classical problem of an introduced pathogen. *Mycologia*, **79**, 23–37.

Andersen, D. C. & MacMahon, J. A. (1981). Population dynamics and bioenergetics of a fossorial herbivore, *Thomomys talpoides* (Rodentia: Geomyidae), in a spruce–fir sere. *Ecological Monographs*, **51**, 179–202.

Anderson, A. (1985). The problems of living with a plant root: factors involved with root colonization. In *Proceedings of the 6th North American Conference on Mycorrhizae*, ed. R. Molina, pp. 175–8. Forestry Sciences Laboratory, Corvallis, OR.

Anderson, A. (1988). Mycorrhizae–host specificity and recognition. *Phytopathology*, **78**, 375–8.

Anderson, R. C. & Liberta, A. (1987). Variation in vesicular-arbuscular mycorrhizal relationships of two sand prairie species. *American Midland Naturalist*, **118**, 56–63.

Anderson, R. C., Liberta, A. E., Dickman, L. A. & Katz, J. A. (1983). Spatial variation in vesicular-arbuscular mycorrhiza spore density. *Bulletin of the Torrey Botanical Club*, **110**, 519–25.

Andrews, H. N. & Lenz, L. Wayne (1943). A mycorrhizome from the carboniferous of Illinois. *Bulletin of the Torrey Botanical Club*, **70**, 120–5.

Anonymous (1931). Establishing Pines. Preliminary observations on the effects of soil inoculation. *Rhodesian Agricultural Journal*, **28**, 185–7.

Antibus, R. K., Kroehler, C. H. & Linkins, A. E. (1986). The effects of external pH, temperature, and substrate concentration on the surface acid phosphatase activity of ectomycorrhizal fungi. *Canadian Journal of Botany*, **64**, 2382–7.

Auge, R. M., Schekel, K. A. & Wample, R. L. (1986). Greater leaf conductance of well-watered VA mycorrhizal plants is not related to phosphorus nutrition. *New Phytologist*, **103**, 107–16.

Azcon, R., Barea, J. M. & Hayman, D. S. (1976). Utilization of rock phosphate in alkaline soils by plants inoculated with mycorrhizal fungi and phosphate-solubilizing bacteria. *Soil Biology and Biochemistry*, **8**, 135–8.

Azcon-Aguilar, C., Gianinazzi, V., Fardeau, J. C. & Gianinazzi, S. (1986). Effect of vesicular-arbuscular mycorrhizal fungi and phosphate-solubilizing bacteria on growth and nutrition of soybean in a neutral–calcareous soil amended with ^{32}P-^{45}Ca-tricalcium phosphate. *Plant and Soil*, **96**, 3–15.

Azcon, R. & Ocampo, J. A. (1981). Factors affecting the vesicular-arbuscular infection and mycorrhizal dependency of thirteen wheat cultivars. *New Phytologist*, **87**, 677–85.

Bagyaraj, D. J. (1984) Biological interactions with VA mycorrhizal fungi. In *VA Mycorrhiza*, ed. C. L. Powell and D. J. Bagyaraj, pp. 131–53. CRC Press, Boca Raton, FL.

Bagyaraj, D. J., Manjunath, A. & Patil, R. B. (1979). Occurrence of vesicular-arbuscular mycorrhizas in some tropical aquatic plants. *Transactions of the British Mycological Society*, **72**, 164–7.

Bajwa, R. & Read, D. J. (1986). Utilization of mineral and amino N sources by the ericoid mycorrhizal endophyte *Hymenoscyphus ericae* and by mycorrhizal and non-mycorrhizal seedlings of *Vaccinium*. *Transactions of the British Mycological Society*, **87**, 269–77.

Barrett, J. A. (1983).Plant–fungus symbioses. In *Coevolution*, ed. D. J. Futuyman & M. Slatkin, pp. 137–60. Sinauer Associates, Inc., Sunderland, MA.

Bartlett, E. M. & Lewis, D. H. (1973). Surface phosphatase activity of mycorrhizal roots of beech. *Soil Biology and Biochemistry*, **5**, 249–57.

Baylis, G. T. S. (1975). The magnolioid mycorrhiza and mycotrophy in root systems derived from it. In *Endomycorrhizas*, ed. F. E. Saunders, B. Mosse & P. B. Tinker, pp. 373–89. Academic Press, London.

Becard, G. & Piche, Y. (1989). Fungal growth stimulation by CO_2 and root exudates in vesicular-arbuscular mycorrhizal symbiosis. *Applied and Environmental Microbiology*, **55**, 2320–5.

Beckjord, P. R. & Hacskaylo, E. (1984). E-strain fungus associations with roots of *Paulownia tomentosa* and *Pinus strobus*. *Bulletin of the Torrey Botanical Club*, **111**, 227–30.

Becker, P. (1983), Ectomycorrhizae on *Shorea* (Dipterocarpaceae) seedlings in a lowland Malaysian rain forest. *Malaysian Forester*, **46**, 146–70.

Bell, A. D. (1984) Dynamic morphology: A contribution to plant population ecology. In *Perpectives on Plant Population Ecology*, ed. R. Dirzo and J. Sarukhan, pp. 48–65. Sinauer Associates Inc., Sunderland, MA.

Belsky, A. J. (1986). Does herbivory benefit plants? A review of the evidence. *American Naturalist*, **127**, 870–92.

Benjamin, P. K. & Allen, E. B. (1987). The influence of VA mycorrhizal fungi on competition between plants of different successional stages in sagebrush-grassland. In *Mycorrhizae in the Next Decade, Practical Applications and Research Priorities*, ed. D. M. Sylvia, L. L. Hung and J. H. Graham, p. 144. IFAS, Gainesville, FL.

Berch, S. M. & Kendrick, B. (1982). Vesicular-arbuscular mycorrhizae of southern Ontario ferns and fern-allies. *Mycologia*, **74**, 769–76.

Berch, S. M. & Warner, B. G. (1985). Fossil vesicular-arbuscular mycorrhizal fungi of two *Glomus* species, Endogonaceae, Zygomycetes from late quaternary deposits in Ontario, Canada. Review of Palaeobotany and Palynology, **45**, 229–38.

Bergen, M. & Koske, R. E. (1984). Vesicular-arbuscular mycorrhizal fungi from sand dunes of Cape Cod, Massachusetts. *Transactions of the British Mycological Society*, **83**, 157–8.

Bernard, N. (1909). L'evolution dans la symbiose. Les archiders et leur champignons commenseaux. *Annales des Sciences Naturelles Botanique*, **9**, 1–196.

Bernard, N. (1911). Les mycorrhizes des Solanum. *Annales des Sciences Naturelles Botanique*, **14**, 235–58.

Bethlenfalvay, G. J., Brown, J. S. & Pacovski, R. S. (1982). Parasitic and mutualistic associations between a mycorrhizal fungus and soybean; development of the host plant. *Phytopathology*, **72**, 889–93.

Bethlenfalvay, G. J. & Dakessian, S. (1984). Grazing effects on mycorrhizal colonization and floristic composition of the vegetation on a semiarid range in Northern Nevada. *Journal of Range Management*, **37**, 312–16.

Bethlenfalvay, G. J., Evans, R. A. & Lesperance, A. L. (1985). Mycorrhizal colonization of crested wheatgrass as influenced by grazing. *Agronomy Journal*, **77**, 233–6.

Bildusas, I. J., Dixon, R. K., Pfleger, F. L. & Steward, E. L. (1986). Growth, nutrition and gas exchange of *Bromus inermis* inoculated with *Glomus fasciculatum*. *New Phytologist*, **102**, 303–11.

Bird, G. W., Rich, J. R. & Glover, S. U., (1974). Increased endomycorrhizae of cotton roots in soil treated with nematicides. *Phytopathology*, **64**, 48–51.

Bjorkman, E. (1944). The effect of strangulation on the formation of mycorrhizae in pine. *Svensk Botanisk Tidskrift*, **38**, 1–14.

Bjorkman, E. (1960). *Monotropa hypopitys* L., an epiparasite on tree roots. *Physiologia Plantarum*, **13**, 308–27.

Bledsoe, C. S. & Rygiewicz, P. T. (1986). Ectomycorrhizas affect ionic balance during ammonium uptake by Douglas-fir roots. *New Phytologist*, **102**, 271–83.

Boerner, R. E. J. (1986). Seasonal nutrient dynamics, nutrient resorption, and mycorrhizal infection intensity in two perennial forest herbs. *American Journal of Botany*, **73**, 1249–57.

Bolan, N. S., Robson, A. D., Burrow, J. J. & Aylmore, L. A. G. (1984). Specific activity of phosphorus in mycorrhizal and nonmycorrhizal plants in relation to the availability of phosphate to all plants. *Soil Biology and Biochemistry*, **16**, 299–304.

Bonfante-Fasolo, P. (1984). Anatomy and morphology of VA mycorrhizae. In *VA Mycorrhiza*, ed. C. L. Powell and D. J. Bagyaraj, pp. 5–33. CRC Press, Boca Raton, Fl.

Bonfante-Fasolo, P., Gianinazzi-Pearson, V. & Martinego, L. (1984). Ultrastructural aspects of endomycorrhiza in the Ericaceae IV. Comparison of infection by *Pezizella ericae* in host and non-host plants. *New Phytologist*, **98**, 329–33.

Bonfante-Fasolo, P. & Perotto, S. (1986). Visualization of surface sugar residues in mycorrhizal ericoid fungi by fluorescein conjugated lectins. *Symbiosis* **1**, 269–88.

Boucher, D. H. (1985). *The Biology of Mutualism, Ecology and Evolution*. Oxford University Press, New York.

Boucher, D. H., James, S. & Keeler, K. H. (1982). The ecology of mutualism. *Annual Review of Ecology and Systematics*, **13**, 315–47.

Bowen, G. D. (1987). The biology and physiology of infection and its development. In *Ecophysiology of VA Mycorrhizal Plants*, ed. G. R. Safir, pp. 27–57. CRC Press, Boca Raton, FL.

Bowen, G. D. & Smith, S. E. (1981). The effects of mycorrhizas on nitrogen uptake by plants. *Ecological Bulletin* (*Stockholm*), **33**, 237–47.

Bradley, R., Burt, A. J. & Read, D. J. (1981) . Mycorrhizal infection and resistance to heavy metal toxicity in *Calluna vulgaris*. *Nature* (*London*), **292**, 335–7.

Bradley, R., Burt, A. J. & Read, D. J. (1982). The biology of mycorrhiza in the Ericaceae. VIII. The role of mycorrhizal infection in heavy metal resistance. *New Phytologist*, **91**, 197–201.

Briscoe, C. B. (1959). Early results of mycorrhizal inoculation of pine in Puerto Rico. *Caribbean Forester*, July–December, 73–7.

Brown, R. T. & Mikola, P. (1974). The influence of fructicose soil lichens upon mycorrhizae and seedling growth of forest trees. *Acta Forestalia Fennica*, **141**, 23.

Brownlee, C. D., Duddridge, J. A., Malibari, A. & Read, D. J. (1983). The structure and function of mycelial systems of ectomycorrhizal roots with special reference to their role in forming inter-plant connections and providing pathways for assimilate and water transport. In *Proceedings, IUFRO Conference on Tree Roots and their Mycorrhizas*, ed. D. Atkinson *et al.*, pp. 433–43. Academic Press, London.

Buchholz, K. & Motto, H. (1981). Abundances and vertical distributions of mycorrhizae in plains and barrens forest soils from the New Jersey Pine Barrents. *Bulletin of the Torrey Botanical Club*, **108**, 268–271.

Buwalda, J. G. & Goh, K. M. (1982). Host–fungus competition for carbon as a cause of growth depressions in vesicular-arbuscular mycorrhizal ryegrass. *Soil Biology and Biochemistry*, **14**, 103–6.

Buwalda, J. G., Ross, G. J. S., Stribley, D. P. & Tinker, P. B. (1982). The development of endomycorrhizal root systems. IV. The mathematical analysis of effects of phosphorus on the spread of vesicular-arbuscular mycorrhizal infection in root systems. *New Phytologist*, **92**, 391–400.

Bye, R. A., Jr (1979). Incipient domestication of mustards in northwest Mexico. *The Kiva*, **44**, 237–45.

Bye, R. A., Jr (1985). Botanical perspectives of Ethnobotany of the Greater Southwest. *Economic Botany*, **39**, 375–86.

Caldwell, M. M. (1976). Root extension and water absorption. In *Ecological Studies, Analysis and Synthesis, Vol. 19, Water and Plant Life*, ed. O. L. Lange, L. Kappen, and E. D. Schulze, pp. 63–85. Springer-Verlag, Berlin.

Caldwell, M. M., Eissenstat, D. M., Richards, J. H. & Allen, M. F. (1985). Competition for phosphorus: differential uptake from dual-isotope-labeled soil interspaces between shrub and grass. *Science*, **229**, 384–6.

Carling, D. E. & Brown, M. F. (1980). Relative effect of vesicular-arbuscular mycorrhizal fungi on the growth and yield of soybeans. *Soil Science Society of America Journal*, **44**, 528–32.

Carpenter, A. T. & Allen, M. F. (1988). Responses of *Hedysarum boreale* to mycorrhizas and *Rhizobium*: plant and soil nutrient changes. *New Phytologist*, **109**, 125–32.

Carpenter, S. E., Trappe, J. M. & Ammirati, J. Jr (1987). Observations of fungal succession in the Mount St. Helens devastation zone, 1980–1983. *Canadian Journal of Botany*, **65**, 716–28.

Castellano, M. A. & Trappe, J. M. (1985). Mycorrhizal associations of five species of Monotropoideae in Oregon. *Mycologia*, **77**, 499–502.

Chakraborty, S., Theodorou, C., & Bowen, G. D. (1985). Mycophagous amoebae reduction of root colonization by *Rhizopogon*. In *Proceedings of the 6th North American Conference on Mycorrhizae*, ed. R. Molina, p. 275. Forestry Sciences Laboratory, Corvallis, OR.

Chapin, F. S. III (1980). The mineral nutrition of wild plants. *Annual Review of Ecology and Systematics*, **11**, 233–60.

Chapin, F. S. III, Bloom, A. J., Field, C. B. & Waring, R. H. (1987). Plant responses to multiple environmental factors. *Bioscience*, **37**, 49–57.

Chapin, F. S. III, Fetcher, N., Kielland, K., Everett, K. & Linkins, A. (1988). Productivity and nutrient cycling of Alaskan tundra: enhancement by flowing soil water. *Ecology*, **69**, 693–702.

Chapin, F. S. III, Oechel, W. C., Van Cleve, K. & Lawrence, W. (1987). The role of mosses in phosphorus cycling of an Alaskan black spruce forest. *Oecologia* (*Berlin*), **74**, 310–15.

Chevalier, G. & Grente, J. (1977). Role of man: planned dispersal of truffle fungi by movement of inoculated hosts. *3rd North American Conference on Mycorrhizae* (abstract).

Chiarello, N., Hickman, J. C. & Mooney, H. A. (1982). Endomycorrhizal role for interspecific transfer of phosphorus in a community of annual plants. *Science*, **217**, 941–3.

Chilvers, G. A., Lapeyrie, F. F. & Horan, D. P. (1987). Ectomycorrhizal vs endomycorrhizal fungi within the same root system. *New Phytologist*, **107**, 441–8.

Christensen, M. (1981). Species diversity and dominance in fungal communities. In *The Fungal Community*, ed. D. T. Wicklow and G. C. Carroll, pp. 201–32. Marcel Dekker, Inc., New York.

Christensen, M. (1989). A view of fungal ecology. *Mycologia*, **81**, 1–19.

Christensen, M. & Allen, M. F. (1980). Effect of VA mycorrhizae on water stress tolerance and hormone balance in native western plant species. 1979 Final Report to RMIEE, Laramie, WY.

Christie, P. & Nicolson, T. N. (1983). Are mycorrhizas absent from the Antarctic? *Transactions of the British Mycological Society*, **80**, 557–61.

Christy, E., Sollins, P. & Trappe, J. M. (1982). First year survival of *Tsuga heterophylla* without mycorrhizae and subsequent ectomycorrhizal development on decaying logs and mineral soil. *Canadian Journal of Botany*, **60**, 1601–5.

Clements, F. E. (1916). Plant succession: an analysis of the development of vegetation. *Carnegie Institute of Washington Publ. No.* **242**, 1–512.

Cluett, H. C. & Boucher, D. H. (1983). Indirect mutualism in the *Legume–Rhizobium*–mycorrhizal fungus interaction. *Oecologia* (*Berlin*), **59**, 405–8.

Coleman, D. C., Reid, C. P. & Cole, C. V. (1983). Biological strategies of nutrient cycling in soil systems. *Advances in Ecological Research*, **13**, 1–55.

Cook, R. J. & Baker, K. F. (1983). The Nature and Practice of Biological Control of Plant Pathogens. American Phytopathological Society, St Paul, MN.

Cooke, J. C., Gemma, J. N. & Koske, R. E. (1987). Observations of nuclei in vesicular-arbuscular mycorrhizal fungi. *Mycologia*, **79**, 331–3.

Cooper, K. M. (1984). Physiology of VA mycorrhizal associations. In *VA Mycorrhiza*, ed. C. L. Powell and D. J. Bagyaraj, pp. 155–86. CRC Press, Boca Raton, FL.

Cowles, H. C. (1901). The physiographic ecology of Chicago and vicinity. *Botanical Gazette*, **31**, 145–82.

Cox, G. & Sanders, F. (1974). Ultrastructure of the host–fungus interface in a vesicular-arbuscular mycorrhiza. *New Phytologist*, **73**, 901–12.

Cox, G. & Tinker, P. B. (1976). Translocation and transfer of nutrients in vesicular-arbuscular mycorrhizas. I. The arbuscule and phosphorus transfer: a quantitative ultrastructural study. *New Phytologist*, **77**, 371–8.

Cress, W. A., Throneberry, G. D. & Lindsey, D. L. (1979). Kinetics of phosphorus absorption by mycorrhizal and nonmycorrhizal tomato roots. *Plant Physiology*, **64**, 484–7.

Cromack, K. Jr (1981). Belowground processes in forest succesion. In *Forest Succession, Concepts and Application*, ed. D. C. West, H. H. Shugart & D. B. Botkin, pp. 361–73. Springer-Verlag, New York.

Cromack, K. Jr, Sollins, K. P., Graustein, W. C., Speidel, K., Todd, A. W., Spycher, G.,

Li, C. Y. & Todd, R. L. (1979). Calcium oxalate accumulation and soil weathering in mats of the hypogeous fungus *Hysterangium crassum*. *Soil Biology and Biochemistry*, **11**, 463–8.
Crowell, H. F. & Boerner, R. E. J. (1988). Influence of mycorrhizae and phosphorus on belowground competition between two old-field annuals. *Environmental and Experimental Botany*, **28**, 381–92.
Crush, J. R. (1974). Plant growth responses to vesicular-arbuscular mycorrhiza in herbage legumes. *New Phytologist*, **73**, 743–9.
Curtis, J. T. (1937). Non-specificity of orchid mycorrhizal fungi. *Proceedings of the Society of Experimental Biology and Medicine*, **36**, 43–4.
Curtis, J. R. (1939). The relation of specificity of orchid mycorrhizal fungi to the problem of symbiosis. *American Journal of Botany*, **26**, 390–8.
Daft, M. J. & Hacskaylo, E. (1976). Arbuscular mycorrhizas in anthracite and bituminous coal wastes of Pennsylvania. *Journal of Applied Ecology*, **13**, 523–31.
Daft, M. J. & Okusanya, B. D. (1973). Effect of *Endogone* on plant growth. *New Phytologist*, **72**, 1333–69.
Daniels-Hetrick, B. A. (1984). Ecology of VA mycorrhizal fungi in *VA Mycorrhiza*, ed. C. L. Powell and D. J. Bagyaraj, pp. 35–55. CRC Press, Boca Raton, FL.
Davidson, D. E. & Christensen, M. (1977). Root–microfungal and mycorrhizal associations in a shortgrass prairie. In: *The Belowground Ecosystem: A Synthesis of Plant-associated Processes*, ed. J. K. Marshall, pp. 279–87. Colorado State University Press, Collins, CO.
Davies, R. M. & Menge, J. A. (1981). *Phytophthora parasitica* inoculation and intensity of vesicular-arbuscular mycorrhizae in citrus. *New Phytologist*, **87**, 705–15.
Dean, A. M. (1983). A simple model of mutualism. *American Naturalist*, **121**, 409–17.
deBary, A. (1887). *Comparative Morphology and Biology of the Fungi, Mycetozoa and Bacteria*. (English translation of the 1884 edition). Clarendon Press, Oxford.
Di, J. J. (1988). Physiological response and VA mycorrhizal infection of several cultivars of Asian *Agropyron* species. MS thesis, Utah State University, Logan, UT.
Di, J. J. & Allen, E. B. (1990). Physiological responses of six wheatgrass cultivars to mycorrhizae. *Journal of Range Management* (in press).
Diehl, W. W. (1939). *Endogone* as animal food. *Science*, **90**, 442.
Dighton, J., Poskitt, J. M. & Howard, D. M. (1986). Changes in occurrence of basidiomycete fruit bodies during forest stand development with specific reference to mycorrhizal species. *Transactions of the British Mycological Society*, **87**, 165–71.
Dimbleby, G. W. (1953). Natural regeneration of pine and birch on the heather moors of northeast Yorkshire. *Forestry*, **26**, 41–52.
Dodd, J. C., Burton, C. C., Burns, R. G. & Jeffries, P. (1987). Phosphatase activity associated with the roots and the rhizosphere of plants infected with vesicular-arbuscular mycorrhizal fungi. *New Phytologist*, **107**, 163–72.
Dominik, T. (1951). Badania mykotrofizmu roslinnosci wydm nadmorschich i srodladowyck. *Acta Societatis Botanica Polonica*, **21**(1–2), 125–64.
Duce, D. H. (1987). Effects of vesicular-arbuscular mycorrhizae on *Agropyron smithii* grown under drought stress and their influence on organic phosphorus mineralization. MS thesis, Utah State University, Logan, UT.
Duddridge, J. A., Finlay, R. D., Read, D. J. & Soderstrom, B. (1988). The structure and function of the vegetative mycelium of ectomycorrhizal plants. III. Ultrastructural and autogradiographic analysis of interplant carbon distribution through intact mycelial systems. *New Phytologist*, **108**, 183–8.
Duddridge, J. A., Malibari, A. & Read, D. J. (1980). Structure and function of mycorrhizal rhizomorphs with special reference to their role in water transport. *Nature (London)*, **287**, 834–6.

Ebbers, B. C., Anderson, R. C. & Liberta, A. E. (1987). Aspects of the mycorrhizal ecology of prairie dropseed, *Sporobolus heterolepis* (Poaceae). *American Journal of Botany*, **74**, 564–73.

Emmons, L. H. (1982). Ecology of Prochimys, Rodentia, Echimyidae, in southeastern Peru. *Tropical Ecology*, **23**, 280–90.

Falck, R. & Falck, M. (1954). *Die Benentung der Fadenpilze als symbiotender pflanzen fur die waldkultur*. J. D. Sauerlanders Verlag, Frankfurt.

Falinski, J. B. (1986). *Vegetation Dynamics in Temperate Lowland Primeval Forests*. Junk Publishers, Netherlands.

Farmer, A. M. (1988). Vesicular-arbuscular mycorrhizae in submerged Isoetes. *The Mycologist*, **2**, 74.

Finlay, R. D. & Read, D. J. (1986). The structure and function of the vegetative mycelium of ectomycorrhizal plants. I. Translocation of ^{14}C-labelled carbon between plants interconnected by a common mycelium. *New Phytologist*, **103**, 143–56.

Fitter, A. H. (1977). Influence of mycorrhizal infection on competition for phosphorus and potassium by two grasses. *New Phytologist*, **79**, 119–25.

Fitter, A. H. (1985a). *Ecological Interactions in Soil; Plants, Microbes and Animals*. British Ecological Society Special Publications, No. R, Blackwell Scientific Publications, Palo Alto, CA.

Fitter, A. H. (1985b). Functioning of vesicular-arbuscular mycorrhizas under field conditions. *New Phytologist*, **99**, 257–65.

Fitter, A. H. (1988). Water relations of red clover *Trifolium pratense* L. as affected by VA mycorrhizal infection and phosphorus supply before and during drought. *Journal of Experimental Botany*, **39**, 595–603.

Fleming, L. V. (1984). Effects of soil trenching and coring on the formation of ectomycorrhizas on birch seedlings grown around mature trees. *New Phytologist*, **98**, 143–53.

Fleming, L. V. (1985). Experimental study of sequences of ectomycorrhizal fungi on birch (*Betula* sp.) seedling root systems. *Soil Biology and Biochemistry*, **17**, 591–600.

Fleming, L. V., Deacon, J. W., Last, F. T. & Donaldson, S. J. (1984). Influence of propagating soil on the mycorrhizal succesion of birch seedlings transplanted to a field site. *Transactions of the British Mycological Society*, **82**, 707–11.

Fogel, R. & Hunt, G. (1979). Fungal and arboreal biomass in a western Oregon Douglas-fir ecosystem: distribution patterns and turnover. *Canadian Journal of Forest Research*, **9**, 245–56.

Fogel, R. & Hunt, G. (1983). Contribution of mycorrhizae and soil fungi to nutrient cycling in a Douglas-fir ecosystem. *Canadian Journal of Forest Research*, **13**, 219–32.

Ford, E. D., Mason, P. A. & Pelham, J. (1980). Spatial patterns of sporophore distribution around a young birch tree in three successive years. *Transactions of the British Mycological Society*, **75**, 287–96.

Forman, R. T. T. & Godron, M. (1986). *Landscape Ecology*. John Wiley & Sons, New York.

Forster, S. M. & Nicolson, T. H. (1981). Aggregation of sand from a maritime embryo sand dune by microorganisms and higher plants. *Soil Biology and Biochemistry*, **13**, 199–203.

Francis, R., Finlay, R. D. & Read, D. J. (1986). Vesicular-arbuscular mycorrhiza in natural vegetation IV. Transfer of nutrients in inter- and intra-specific combinations of host plants. *New Phytologist*, **102**, 103–11.

Francl, L. J. & Dropkin, V. H. (1985). *Glomus fasciculatum*, a weak pathogen of *Heterodera glycines*. *Journal of Nematology*, **17**, 470–5.

Frank, A. B. (1885). Ueber die auf Wurzelsymbiose beruhende Ernährung gewisser Baume durch unterirdische Pilze. *Berichte der Deutsche Botanische Gesellschaft*, **3**, 128–45.

Frank, A. B. (1887). Ueber neue Mykorrhiza-formen. *Berichte der Deutsche Botanische Gesellschaft*, **5**, 395–422.
Frank, A. B. (1888). Ueber die physiologische Bedeutung der Mycorrhiza. *Berichte der Deutsche Botanische Gesellschaft*, **6**, 248–68.
Frank, A. B. (1891). Ueber die auf Verdauung von Pilzen abzielende Symbiose der mit endotrophen mykorrhizen begabten Pflanzen, sowie der Leguminosen und Erlen. *Berichte de Deutsche Botanische Gesellschaft*, **9**, 244–53.
Frank, A. B. (1894). Die Bedeutung det Mykorrhiza-pilze fur die gemeine Kiefer. *Forstwissenschaaftliches*, **16**, 1852–90.
Freckman, D. W., Allen, M. F. & Wharton, R. A. Jr (1988). *Nematodes and VA mycorrhizae in the Antarctic Dry Valley, Exobiology and Future Mars Missions*. NASA, Ames, Mountain View, CA (abstract).
Fries, N.(1987). Somatic incompatibility and field distribution of the ectomycorrhizal fungus *Suillus luteus* (Bolataceae). *New Phytologist*, **107**, 735–9.
Fries, N. & Birraux, D. (1980). Spore germination in *Hebeloma* stimulated by living plant roots. *Experientia*, **36**, 1056–7.
Friese, C. F. (1984). The distribution of VAM fungi in a sand dune. MS thesis, University of Rhode Island, Kingston, RI.
Friese, C. F. & Allen, M. F. (1988). The interaction of harvester ant activity and VA mycorrhizal fungi. *Proceedings of the Royal Society of Edinburgh*, **94B**, 176.
Friese, C. F. & Allen, M. F. (1990). Tracking the fates of exotic and local VA mycorrhizal fungi: Methods and patterns. *Agriculture, Ecosystems and Environments (in press)*.
Frydman, I. (1957). Mykotrofizm roslinnoscie pokrywajecej gruzy ruiny domow wroclawia. *Acta Societates Botanica Polonica*, **26**, 45–60.
Gadgil, R. L. & Gadgil, P. D. (1971). Mycorrhiza and litter decomposition. *Nature (London)*, **233**, 133.
Gadgil, R. L. & Gadgil, P. D. (1975). Suppression of litter decomposition by mycorrhizal roots of *Pinus radiata. New Zealand Journal of Forest Science*, **5**, 35–41.
Gardner, I. C., Clelland, D. M. & Scott, A. (1984). Mycorrhizal improvement in non-leguminous nitrogen fixing associations with particular reference to *Hippophae rhamnoides* L. *Plant and Soil*, **78**, 189–99.
Garrett, S. D. (1970). *Pathogenic Root-infecting Fungi*. Cambridge University Press, Cambridge.
Gemma, J. N. & Koske, R. E. (1988). Pre-infection interactions between roots and the mycorrhizal fungus *Gigaspora gigantea*: chemotropism of germ-tubes and root growth response. *Transactions of the British Mycological Society*, **91**, 123–32.
Gerdemann, J. W. (1968). Vesicular-arbuscular mycorrhiza and plant growth. *Annual Review of Phytopathology*, **6**, 397–418.
Gerdemann, J. W. (1974). Vesicular-arbuscular mycorrhizae. In *The development and function of roots*, ed. J. G. Torrey and D. T. Clarkson, pp. 575–91. Academic Press, New York.
Gerdemann, J. W. & Trappe, J. M. (1974). Endogonaceae of the Pacific Northwest. *Mycologia Memoir*, **5**, 1–76.
Gianinazzi-Pearson, V. & Gianinazzi, S. (1983). The physiology of vesicular-arbuscular mycorrhizal roots. *Plant and Soil*, **71**, 197–209.
Gibson, D. J. & Hetrick, B. A. Daniels (1988). Topographic and fire effects on the composition and abundance of VA-mycorrhizal fungi in tallgrass prairie. *Mycologia*, **80**, 433–41.
Gildon, A. & Tinker, P. B. (1981). A heavy metal tolerant strain of a mycorrhizal fungus. *Transactions of the British Mycological Society*, **77**, 648–9.
Gleason, H. A. (1926). The individualistic concept of the plant association. *Bulletin of the Torrey Botanical Club*, **53**, 7–26.

Glenn, M. G., Chew, F. S. & Williams, P. H. (1985). Hyphal penetration of *Brassica* (Cruciferae) roots by a vesicular-arbuscular mycorrhizal fungus. *New Phytologist*, **99**, 463–72.

Glenn, M. G., Chew, F. S. & Williams, P. H. (1988). Influence of glucosinolate content on *Brassica* (Cruciferae) roots on growth of vesicular-arbuscular mycorrhizal fungi. *New Phytologist*, **110**, 217–225.

Goss, R. W. (1960). Mycorrhizae of Ponderosa Pine in Nebraska grassland soils. *University of Nebraska, College of Agriculture Research Bulletin*, **192**, 1–47.

Grace, J. B. & Tilman, G. D. (1990). *Perspectives on Plant Competition*. Academic Press, New York.

Graham, J. H. (1987). Non-nutritional benefits of VAM fungi-do they exist? In *Mycorrhizae in the Next Decade, Practical Applications and Research Priorities*, ed. D. M. Sylvia, L. L. Hung and J. H. Graham, pp. 237–9. IFAS, Gainesville, FL.

Graham, J. H., Leonard, R. T. & Menge, J. A. (1981). Membrane-mediated decrease in root exudation responsible for phosphorus inhibition of vesicular-arbuscular mycorrhiza formation. *Plant Physiology*, **68**, 548–52.

Graham, J. H., Linderman, R. C. & Menge, J. A. (1982). Development of external hyphae by different isolates of mycorrhizal *Glomus* spp. in relation to root colonization and growth of Troyer citrange. *New Phytologist*, **91**, 183–9.

Graham, J. H. & Menge, J. A. (1982). Influence of vesicular-arbuscular mycorrhizae and soil phosphorus on take-all disease of wheat. *Phytopathology*, **72**, 95–8.

Graham, J. H., Syvertsen & Smith, M. L., Jr (1987). Water relations of mycorrhizal and phosphorus-fertilized non-mycorrhizal *Citrus* under drought stress. *New Phytologist*, **105**, 411–19.

Graustein, V. C., Cromack, K. Jr & Sollins, P. (1977). Calcium oxalate: Occurrence in soils and effect on nutrient geochemical cycles. *Science*, **198**, 1252–4.

Grime, J. P. (1979). *Plant Strategies and Vegetation Processes*. John Wiley & Sons, New York.

Grime, J. P., Mackey, J. M. L., Hillier, S. H. & Read, D. J. (1987). Floristic diversity in a model system using experimental microcosms. *Nature* (*London*), **328**, 420–2.

Hacskaylo, E. (1967). Mycorrhizae: indespensable invasions by fungi. *Agricultural Sciences Review*, **5**, 1–36.

Hadley, G. (1984). Uptake of [^{14}C] glucose by a symbiotic and mycorrhizal orchid protocorms. *New Phytologist*, **96**, 263–73.

Hadley, G. (1985). Mycorrhiza in tropical orchids. In *Proceedings of the Fifth Asian Orchid Congress Seminar*, pp. 154–9.

Hall, I. R. (1978). Effects of endomycorrhizas on the competitive ability of white clover. *New Zealand Journal of Agricultural Research*, **21**, 509–15.

Hall, I. R. (1979). Effect of vesicular-arbuscular mycorrhizas on growth of white clover, lotus, and ryegrass in some eroded soils. *New Zealand Journal of Agricultural Research*, **22**, 479–84.

Halling, R. E. & Ovrebo, C. L. (1987). A new species of Rozites from oak forests of Colombia, with notes on Biogeography. *Mycologia*, **79**, 674–8.

Hardie, K. (1985). The effect of removal of extraradical hyphae on water uptake by vesicular-arbuscular mycorrhizal plants. *New Phytologist*, **101**, 677–84.

Harlan, J. R. (1975). *Crops and Man*. American Society of Agronomy, Madison, WI.

Harley, J. L. (1968). Presidential address, Fungal Symbiosis. *Transactions of the British Mycological Society*, **51**, 1–11.

Harley, J. L. (1969). *The Biology of Mycorrhizae*, 2nd edition. Leonard Hill, London.

Harley, J. L. (1971). Fungi in ecosystems. *Journal of Applied Ecology*, **8**, 627–42.

Harley, J. L. & Smith, S. E. (1983). *Mycorrhizal Symbiosis*. Academic Press, London.

Harper, J. L. (1977). *Population Biology of Plants*. Academic Press, London.

Harris, D. & Paul, E. A. (1987). Carbon requirements of vesicular-arbuscular mycorrhizae. In *Ecophysiology of VA Mycorrhizal Plants*, ed. G. R. Safir, pp. 93–105. CRC Press, Boca Raton, FL.

Hartig, T. (1840). *Voll standige Naturgeschichte der forstlichen Culturpflanzen Deutschlands*. A. Forstner'sche Verlagbuchhandlung, Berlin.

Harvey, A. E., Jurgensen, M. F. & Larsen, M. J. (1978). Seasonal distribution of ectomycorrhizae in a mature Douglas-fir/Larch forest soil in western Montana. *Forest Science*, **24**, 203–8.

Harvey, A. E., Jurgensen, M. F. & Larsen, M J. (1980). Clearcut harvesting and ectomycorrhizae: survival of activity on residual roots and influence on a bordering forest stand in western Montana. *Canadian Journal of Forest Research*, **10**, 300–3.

Haselwandter, K., Bonn, G. & Read, D. J. (1987). Degradation and utilization of lignin by mycorrhizal fungi. In *Mycorrhizae in the Next Decade, Practical Applications and Research Priorities*, ed. D. M. Sylvia, L. L. Hung and J. H. Graham, p. 131. IFAS, Gainesville, FL.

Haselwandter, K. & Read, D. J. (1980). Fungal associations of roots of dominant and subdominant plants in high alpine vegetation systems with special reference to mycorrhizae. *Oecologia (Berlin)*, **45**, 57–62.

Hatch, A. B. (1936). The role of mycorrhizae in afforestation. *Journal of Forestry*, **34**, 22–9.

Hatch, A. B. (1937). The physical basis of mycotrophy in *Pinus*. *The Black Rock Forest Bulletin*, **6**, 1–168.

Hattingh, M. J., Gray, L. E. & Gerdemann, J. W. (1973). Uptake and translocation of ^{32}P-labelled phosphate to onion roots by endomycorrhizal fungi. *Soil Science*, **116**, 383–7.

Hayman, D. S. (1987). VA mycorrhizas in field crop systems. In *Ecophysiology of VA Mycorrhizal Plants* ed. G. R. Safir, pp. 171–92. CRC Press, Boca Raton, FL.

Heithaus, E. R., Culver, D. C. & Beattie, A. J. (1980). Models of some ant–plant mutualisms. *American Naturalist*, **116**, 347–61.

Hendrix, L. B. & Smith, S. D. (1986). Post-eruption revegetation of Isla Fernandine, Galapagos: II. *National Geographic Research*, **2**, 6–16.

Henriksson, E. & Henriksson, L. E. (1988). Fungi in Surtsey soils. *Proceedings of the Royal Society of Edinburgh*, **94B**, 61.

Hepper, C. M. (1983). Limited independent growth of a vesicular-arbuscular mycorrhizal fungus *in vitro*. *New Phytologist*, **93**, 537–42.

Hepper, C. M., Sen, R. & Maskall, C. S. (1986). Identification of vesicular-arbuscular mycorrhizal fungi in roots of leek (*Allium porrum* L.) and maize (*Zea mays* L.) on the basis of enzyme mobility during polyacrylamide gel electrophoresis. *New Phytologist*, **102**, 529–39.

Hepper, C. M., Azcon-Aguilar, C., Rosendahl, S. & Sne, R. (1988). Competition between three species of *Glomus* used as spatially separated introduced and indigenous mycorrhizal inocula for leek (*Allium porrum* L.). *New Phytologist*, **110**, 207–15.

Herrera, R., Merida, T., Stark, N. & Jordan, C. F. (1978). Direct phosphorus transfer from leaf litter to roots. *Naturwissenschaften*, **65**, 208–9.

Hetrick, B. A. Daniels (1984). Ecology of VA mycorrhizal fungi. In *VA Mycorrhiza*, ed. C. L. Powell and D. J. Bagyaraj, pp. 35–55. CRC Press, Boca Raton, FL.

Hetrick, B. A. Daniels, Leslie, J. F. & Wilson, G. T. (1988). Physical and topological assessment of effects of a vesicular-arbuscular mycorrhizal fungus on root architecture of big bluestem. *New Phytologist*, **110**, 85–96.

Hinckley, T. W., Smith, W. K. & Christensen, M. (1989). Infectivity and effectivity of indigenous vesicular-arbuscular mycorrhizal fungi from contiguous soils in south-western Wyoming, USA. *New Phytologist*, **112**, 205–14.

Hirrel, M. C. & Gerdemann, J. W. (1979). Carbon transfer between onions infected with vesicular-arbuscular mycorrhizal fungus. *New Phytologist*, **83**, 731–8.

Hirrel, M. C., Mehravaran, H. & Gerdemann, J. W. (1978). Vesicular-arbuscular mycorrhizae in the Chenopodiaceae and Cruciferae: do they occur? *Canadian Journal of Botany*, **56**, 2813–17.

Ho, I. & Trappe, J. M. (1973). Translocation of ^{14}C from *Festuca* plants to their endomycorrhizal fungi. *Nature New Biology*, **244**, 30–1.

Ho, I. & Trappe, J. M. (1975). Nitrate reducing capacity of two vesicular-arbuscular mycorrhizal fungi. *Mycologia*, **67**, 886–8.

Ho, I. & Zak, B. (1979). Acid phosphatase activity of six ectomycorrhizal fungi. *Canadian Journal of Botany*, **57**, 1203–5.

Hoffman, M. T. & Mitchell, O. T. (1986). The root morphology of some legume species in the Southwestern Cape and the relationship of vesicular-arbuscular mycorrhizas with dry mass and phosphorus content of *Acacia saligna* seedlings. *South African Journal of Botany*, **52**, 316–20.

Högberg, P. (1982). Mycorrhizal associations in some woodland and forest trees and shrubs in Tanzania. *New Phytologist*, **92**, 407–15.

Horak, E. (1983). Mycogeography in the South Pacific Region, Agaricales, Boletales. *Australian Journal of Botany* (Supplemental), **10**, 1–42.

Huang, R. S., Smith, W. K. & Yost, R. S. (1985). Influence of vesicular-arbuscular mycorrhiza on growth, water relations, and leaf orientation in *Leucaena leucocephala* (Lam) de Wit. *New Phytologist*, **99**, 229–44.

Iqbal, S. H. & Qureshi, K. S. (1976). The influence of mixed sowing (cereals and crucifers) and crop rotation on the development of mycorrhiza and subsequent growth of crops under field conditions. *Biologic* (*Lahore*) **22**, 287–98.

Ingham, E. R., Trofymow, J. A., Ames, R. N., Hunt, H. W., Morley, C. R., Moore, J. C., & Coleman, D. C. (1986). Trophic interactions and nitrogen cycling in a semi-arid grassland soil. II. System responses to removal of different groups of soil microbes or fauna. *Journal of Applied Ecology*, **23**, 615–30.

Iyer, J. G., Corey, R. B. & Wilde, S. A. (1980). Mycorrhizae: facts and fallacies. *Journal of Arboriculture*, **6**, 213–20.

Jabaji-Hare, S. (1988). Lipid and fatty acid profiles of some vesicular-arbuscular mycorrhizal fungi: contribution to taxonomy. *Mycologia*, **80**, 622–9.

Jakobsen, I. & Nielsen, N. E. (1983). Vesicular-arbuscular mycorrhizae infection in cereals and peas at various times and soil depths. *New Phytologist*, **93**, 401–13.

Janos, D. P. (1980). Mycorrhizae influence tropical succession. *Biotropica*, **12**, 56–64.

Janos, D. P. (1981). V-A mycorrhizae increase productivity and diversity of tropical tree communities. *Fifth North American Conference on Mycorrhizae*. Quebec, Canada p. 18 (abstract).

Janos, D. P.(1987). VA mycorrhizas in humid tropical ecosystems. In *Ecophysiology of VA Mycorrhizal Plants*, ed. G. R. Safir, pp. 107–34. CRC Press, Boca Raton, FL.

Jeffrey, C. (1962). The origin and differentiation of the archegoniate land-plants. *Botaniska Notiser*, **115**, 446–54.

Johnson, P. N. (1976). Effects of soil phosphate level and shade on plant growth and mycorrhizas. *New Zealand Journal of Botany*, **14**, 333–40.

Jurinak, J. J., Dudley, L. M., Allen, M. F. & Knight, W. G. (1986). The role of calcium oxalate in the availability of phosphorus in soils of semiarid regions: a thermodynamic study. *Soil Science*, **142**, 255–61.

Kamienski, F. (1882) (Transl. by S. M. Berch, 1985). The vegetative organs of *Monotropa hypopitys* L. In *Proceedings of the 6th North American Conference on Mycorrhizae*, ed. R. Molina, pp. 12–17. Forest Research Laboratory, Corvallis, OR.

Kelly, A. P. (1950). *Mycotrophy in Plants*. Chronica Botanica Company, Waltham, MA.

Kessell, S. L. (1927). Soil organisms. The dependence of certain pine species on a biological soil factor. *Empire Forestry Journal*, **6**, 70–4.

Khan, A. G. (1972). The effect of mycorrhizal associations on growth of cereals. I. Effects of maize growth. *New Phytologist*, **71**, 613–19.

Khan, A. G. (1974). The occurrence of mycorrhizas in halophytes, hydrophytes and xerophytes and of *Endogone* spores in adjacent soils. *Journal of General Microbiology*, **81**, 7–14.

Khan, A. G. (1975). The effect of vesicular-arbuscular mycorrhizal associations on growth of cereals. II. Effects on wheat growth. *Annals of Applied Biology*, **80**, 27–36.

Khan, A. G. (1978). Vesicular-arbuscular mycorrhiza in plants colonizing black wastes from bituminous coal mining in the Illawarra region of New South Wales. *New Phytologist*, **81**, 53–63.

Khudairi, A. K. (1969). Mycorrhiza in desert soils. *BioScience*, **19**, 598–9.

Kidstone, R. & Lang, W. H. (1921). On the old red sandstone plants showing structure from the Rhynie chart bed Aberdeenshire. Part V. The thallophyte occurring in the conditions of accumulation and preservation of the deposit. *Transactions of the Royal Society of Edinburgh*, **52**, 855–902.

Klopatek, C. C., DeBano, L. F. & Klopatek, J. M. (1988). Effects of simulated fire on vesicular-arbuscular mycorrhizae in pinyon–juniper woodland soil. *Plant and Soil*, **109**, 245–9.

Knight, W. G., Allen, M. F., Jurinak, J. J. & Dudley, L. M. (1989). Elevated carbon dioxide and solution phosphorus in soil with vesicular-arbuscular mycorrhizal western wheatgrass. *Soil Science Society of America Journal*, **53**, 1075–82.

Knutson, D. M., Hutchins, A. S. & Cromack, K. (1980). The association of calcium oxalate-utilizing *Streptomyces* with conifer ectomycorrhizae. *Antonie Van Leewenhoek Journal of Microbiology and Serology*, **46**, 611–19.

Koh, S. D. & Lee, H. H. (1984). Studies of species and distribution of vesicular-arbuscular mycorrhizal fungi in relation to salt-marsh plants. *Korean Journal of Mycology*, **12**, 175–82.

Koide, R. (1985). The nature of growth depressions in sunflower caused by vesicular-arbuscular mycorrhizal infection. *New Phytologist*, **99**, 449–62.

Koide, R., Mingguang, L., Lewis, J. & Irby, C. (1988). Role of mycorrhizal infection in the growth and reproduction of wild versus cultivated plants. 1. Wild vs. cultivated oats. *Oecologia* (*Berlin*), **77**, 537–43.

Koide, R. T. & Mooney, H. A. (1987). Spatial variation in inoculum potential of vesicular-arbuscular mycorrhizal fungi caused by formation of gopher mounds. *New Phytologist*, **107**, 173–82.

Kormanik, P. P. (1985). Effects of phosphorus and vesicular-arbuscular mycorrhizae on growth and leaf retention of black walnut, *Juglans nigra*, seedlings. *Canadian Journal of Forest Research*, **15**, 688–93.

Kormanik, P. P., Bryan, W. C. & Schultz, R. C. (1980). Increasing endomycorrhizal fungus inoculum in forest nursery soil with cover crops. *Southern Journal of Applied Forestry*, **4**, 151–3.

Koske, R. E. (1981). A preliminary study of interactions between species of vesicular-arbuscular fungi in a sand dune. *Transactions of the British Mycological Society*, **76**, 411–16.

Koske, R. E. (1982). Evidence for a volatile attractant from plant roots affecting germ tubes of a VA mycorrhizal fungus. *Transactions of the British Mycological Society*, **79**, 305–10.

Koske, R. E. (1984). Spores of VAM fungi inside spores of VAM fungi. *Mycologia*, **76**, 853–62.

Koske, R. E. (1987). Distribution of VA mycorrhizal fungi along a latitudinal temperature gradient. *Mycologia*, **79**, 55–68.

Koske, R. E., Friese, C. F., Olexia, P. D. & Hauke, R. L. (1985). Vesicular-arbuscular mycorrhizas in *Equisetum*. *Transactions of the British Mycological Society*, **85**, 350–3.

Koske, R. E. & Halvorson, W. L. (1981). Ecological studies at vesicular-arbuscular mycorrhizae in a barrier sand dune. *Canadian Journal of Botany*, **59**, 1413–22.

Koske, R. E. & Polson, W. R. (1984). Are VA mycorrhizae required for sand dune stabilization? *BioScience*, **34**, 420–4.

Koslowski, S. D. & Boerner, R. E. J. (1989). Interactive effects of aluminium, phosphorus, and mycorrhizae on growth and nutrient uptake of *Panicum virgatum* L. (Poaceae). *Environmental Pollution* (in press).

Kotter, M. M. & Farentinos, R. C. (1984). Tassel-eared squirrels *Sciurus aberti ferreus* as spore dispersal agents of hypogeous mycorrhizal fungi. *Journal of Mammalogy*, **65**, 684–7.

Kough, J., Malajczuk, N. & Lindermann, R. (1983). Use of the indirect immunofluorescent technique to study the vesicular-arbuscular fungus, *Glomus epigaeum* and other *Glomus* species. *New Phytologist*, **94**, 57–62.

Koviacic, D. A., St John, T. V. & Dyer, M. I. (1984). Lack of vesicular-arbuscular mycorrhizal inoculum in a ponderosa pine *Pinus ponderosa* forest. *Ecology*, **65**, 1755–9.

Kowalski, S. (1977). An attempt of estimating the effect of fungal communities in the forest soil environment on the fungal components of scots pine (*Pinus silvestris* L.) mycorrhizae. *Acta Agraria et Silvestria*, **17**, 51–63.

Kowalski, S. (1980). Influence of soil fungus community in selected mountain stands on the development of *Cylindrocarpum destructans* (Zins.) Scholt. *Acta Societatis Botanica Polonica*, **49**, 487–92.

Kowalski, S. (1982). Role of mycorrhiza and soil fungi in natural regeneration of fir (*Abies alba Mill*) in Polish Carpathians and Sudetes. *European Journal of Forest Pathology*, **12**, 107–12.

Kowalski, S. (1987). Mycotrophy of trees in converted stands remaining under strong pressure of industrial pollution. *Angew. Botanik*, **61**, 65–83.

Kramer, P. J. & Wilbur, K. W. (1949). Absorption of radioactive phosphorus by mycorrhizal roots of pine. *Science*, **110**, 8–9.

Krishna, K. R., Suresh, H. M., Syamsunder, J. & Bagyaraj, D. J. (1981). Changes in the leaves of fingermillet due to VA mycorrhizal infection. *New Phytologist*, **87**, 717–22.

Kroehler, C. J., Antibus, R. K. & Linkins, A. E. (1988). The effects of organic and inorganic phosphorus concentration on the acid phosphatase activity of ectomycorrhizal fungi. *Canadian Journal of Botany*, **66**, 750–6.

Kucey, R. M. N. & Paul, E. A. (1982a). Carbon flow, photosynthesis, and N_2 fixation in mycorrhizal and nodulated faba beans (*Vicia faba* L.). *Soil Biology and Biochemistry*, **14**, 407–12.

Kucey, R. M. N. & Paul, E. A. (1982b). Biomass of mycorrhizal fungi associated with bean roots. *Soil Biology and Biochemistry*, **14**, 413–14.

Lapeyrie, F. F. (1988). Oxalate synthesis from soil bicarbonate by the mycorrhizal fungus *Paxillus involutus*. *Plant and Soil*, **110**, 3–8.

Lapeyrie, F. F. & Bruchet, (1986). Calcium accumulation by two strains, calcicole and calcifuge, by the mycorrhizal fungus *Paxillus involutus*. *New Phytologist*, **103**, 133–41.

Lapeyrie, F. F. & Chilvers, G. A. (1985). An endomycorrhiza–ectomycorrhiza succession associated with enhanced growth of *Eucalyptus dumosa* seedlings planted in a calcareous soil. *New Phytologist*, **100**, 93–104.

Laursen, G. A. (1985). Mycorrhizae: a review of the importance of fungi from high latitude forests of Alaska, USA. *Agroborealis*, **17**, 58–66.

Lesica, P. & Antibus, R. K. (1986). Mycorrhizal status of hemiparasitic vascular plants in Montana, USA. *Transactions of the British Mycological Society*, **86**, 341–3.

Levy, Y. & Krikun, J. (1979). Effect of vesicular-arbuscular mycorrhiza in *Citrus jambhir* water relations. *New Phytologist*, **85**, 25–32.

Lewis, D. H. (1973). Concepts in fungal nutrition and the origin of biotrophy. *Biological Reviews*, **48**, 261–78.

Lewis, D. H. & Harley, J. L. (1965). Carbohydrate physiology of mycorrhizal roots of beech. III. Movement of sugars between host and fungus. *New Phytologist*, **64**, 256–69.

Lindsey, D. L. (1984). The role of vesicular-arbuscular mycorrhizae in shrub establishment. In *VA Mycorrhizae and Regulation of Arid and Semi-Arid Lands*, ed. S. E. Williams and M. F. Allen, pp. 53–68. University of Wyoming Agricultural Experiment Station, Laramie, WY.

LoBuglio, I. F. & Wilcox, H. E., (1988). Growth and survival of ectomycorrhizal and ectendomycorrhizal seedlings of *Pinus resinosa* on iron tailings. *Canadian Journal of Botany*, **66**, 55–60.

Loehle, C. (1988). Problems with the triangular model for representing plant strategies. *Ecology*, **69**, 284–6.

Lohman, M. L. (1927). Occurrence of mycorrhiza in Iowa forest plants. *Univ. of Iowa, Studies in Natural History*, **11**, 33–8.

MacMahon, J. A. (1981). Successional processes: Comparisons among biomes with special reference to probable roles of and influences on animals. In *Forest Succession. Concept and Application*, ed. H. Shugart, D. Botkin and D. West, pp. 277–304. Springer Verlag, New York.

MacMahon, J. A., Phillips, D. L., Robinson, J. V. & Schimpf, D. J. (1978). Levels of biological organization: an organism-centered approach. *BioScience*, **28**, 700–4.

MacMahon, J. A., Schimpf, D. J., Andersen, D. C., Smith, K. G. & Bayn, R. L. Jr (1981). An organism-centered approach to some community and ecosystem concepts. *Journal of Theoretial Biology*, **88**, 287–307.

MacMahon, J. A. & Warner, N. (1984). Dispersal of mycorrhizal fungi: Processes and agents. In *VA Mycorrhizae and Reclamation of Arid and Semiarid Lands*, ed. S. E. Williams and M. F. Allen, pp. 28–41. University of Wyoming Agricultural Experiment Station, Laramie, WY.

Mainero, J. S. & del Rio, C. M. (1985). Cheating and taking advantage in mutualistic associations. In *The Biology of Mutualism, Ecology and Evolution*, ed. D. H. Boucher, pp. 192–216. Oxford University Press, New York.

Malajczuk, N., Molina, R. & Trappe, J. M. (1984). Ectomycorrhiza formation in Eucalyptus. II. The ultrastructure of compatible and incompatible mycorrhizal fungi and associated roots. *New Phytologist*, **96**, 43–53.

Malajczuk, N. Trappe, J. M. & Molina, R. (1987). Interrelationships among some ectomycorrhizal trees, hypogeous fungi and small mammals: Western Australian and Northwestern American parallels. *Australian Journal of Ecology*, **12**, 53–5.

Malloch, D. & Malloch, B. (1981). The mycorrhizal status of boreal plants: species from Northeastern Ontario. *Canadian Journal of Botany*, **59**, 2167–72.

Malloch, D. & Malloch, B. (1982). The mycorrhizal status of boreal plants: additional species from Northeastern Ontario. *Canadian Journal of Botany*, **60**, 1035–40.

Malloch, D. W., Pirozynski, K. A. & Raven, P. H. (1980). Ecological and evolutionary significance of mycorrhizal symbioses in vascular plants (a review). *Proceedings of the National Academy of Sciences, USA*, **77**, 2113–8.

Margulis, L. & Bermudes, D. (1985). Symbiosis as a mechanism of evolution: status of cell symbiosis theory. *Symbiosis*, **1**, 101–124.

Marks, G. C. & Foster, R. C., (1967). Succession of mycorrhizal associations on individual roots of radiata pine. *Australian Forestry*, **31**, 94–201.

Marks, G. C. & Kozlowski, T. T. (ed.) (1973). *Ectomycorrhizae; Their Structure and Function*. Academic Press, New York.

Martin, F., Stewart, G. R., Genetet, I. & LeTacon, F. (1986). Assimilation of $^{15}NH_4^+$ by beech (*Fagus sylvatica* L.) ectomycorrhizas. *New Phytologist*, **102**, 85–94.
Marx, D. H. (1969a). Antagonism of mycorrhizal fungi to root pathogenic fungi and soil bacteria. *Phytopathology*, **56**, 53–163.
Marx, D. H. (1969b). Production, identification, and biological activity of antibiotics produced by *Leucopaxillus ceralis* var. *piceina*. *Phytopathology*, **59**, 411–17.
Marx, D. H. (1975). Mycorrhizae and establishment of trees on strip mined lands. *The Ohio Journal of Science*, **75**, 288–97.
Marx, D. H. (1977). Tree host range & world distribution of the ectomycorrhizal fungus *Pisolithus tinctorius*. *Canadian Journal of Microbiology*, **23**, 217–23.
Marx, D. H., Cordell, C. E., Kenney, D. S., Mexal, J. G., Artman, J. D., Riffle, J. W. & Molina, R. J. (1984). Commercial vegetative inoculum of *Pisolithus tinctorius* and inoculation techniques for development of ectomycorrhizae on bare-root seedlings. *Forest Science Monograph*, **25**, 1–101.
Marx, D. H. & Davey, C. B. (1969a). Resistance of aseptically formed mycorrhizae to infection by *Phytophthora cinnamoni*. *Phytopathology*, **59**, 549–58.
Marx, D. H. & Davey, C. B. (1969b). Resistance of naturally occurring mycorrhizae to infections by *Phytophthora cinnamoni*. *Phytopathology*, **59**, 559–65.
Maser, C. (1988). Of time and the forest. Buried Treasures. *Natural History*, **97**, 58–59.
Maser, C., Maser, Z. & Molina, R. (1988). Small-mammal mycophagy in rangelands of central and southeastern Oregon. *Journal of Range Management*, **41**, 309–12.
Maser, C., Trappe, J. M. & Nussbaum, R. A. (1978a). Fungal-small mammal interrelationships with emphasis on Oregon forests. *Ecology*, **59**, 799–809.
Maser, C., Trappe, J. M. & Ure, D. C. (1978b). Implications of small mammal mycophagy to the management of western coniferous forests. *Transactions of the 43rd North American Wildlife and Natural Resources Conference*, pp. 78–88.
May, R. M. (1974). *Stability and Complexity in Model Ecosystems*. Princeton University Press, Princeton, NJ.
May, R. M. (1981). Models for two interacting populations. In *Theoretical Ecology, Principles and Applications*, ed. R. M. May, pp. 78–104. Sinauer Associates, Sunderland, MA.
McDougall, W. B. (1918). The classification of symbiotic phenomena, *Plant World*, **21**, 250–6.
McDougall, W. B. & Liebtag, C. (1928). Symbiosis in a deciduous forest. III. Mycorrhizal relations. *Botanical Gazette*, **86**, 226–34.
McGonigle, T. P. & Fitter, A. H. (1988). Ecological consequences of arthropod grazing on VA mycorrhizal fungi. *Proceedings of the Royal Society of Edinburgh*, **94B**, 25–32.
McIlveen, W. D. & Cole, H. Jr (1976). Spore dispersal of endogonaceae by worms, ants, wasps and birds. *Canadian Journal of Botany*, **54**, 1486–9.
McIntosh, R. P. (1985). *The Background of Ecology, Concept and Application*. Cambridge University Press, Cambridge.
McNaughton, J. S. (1983). Serengeti grassland ecology: the role of composite environmental factors and contingency in community organization. *Ecological Monographs*, **53**, 291–320.
Mejstrik, V. (1984). Ecology of vesicular-arbuscular mycorrhizae of the schoenetum-nigricantis-bohenicum community in the Grabanovsky swamps reserve USSR. *Soviet Journal of Ecology*, **15**, 18–23.
Melin, E. (1953). Physiology of mycorrhizal relations in plants. *Annual Review of Plant Physiology*, **4**, 325–46.
Melin, E. & Nilsson, H. (1953). Transfer of labelled nitrogen from glutamic acid to pine seedlings through the mycelium of *Boletus variegatus* (S.W.) Fr. *Nature* (*London*), **171**, 434.

Menge, J. A. & Grand, L. F. (1981). Mycorrhizae of eleven year old *Pinus taeda* plantations in North Carolina, USA. *Journal of the Elish Mitchell Science Society*, **97**, 55–66.

Merrell, D. J. (1981). *Ecological Genetics*. University of Minnesota Press, Minneapolis, MN.

Mexal, J. & Reid, C. P. P. (1973). The growth of selected mycorrhizal fungi in response to induced water stress. *Canadian Journal of Botany*, **51**, 1579–88.

Meyer, F. H. (1973). Distribution of ectomycorrhizae in native and man-made forests. In *Ectomycorrhizae: their ecology and physiology*, ed. G. C. Marks and T. T. Kozlowski, pp. 79–105. Academic Press, New York.

Mikola, P. (1953). An experiment on the invasion of mycorrhizal fungi into prairie soil. *Karstenia*, **2**, 33–4.

Mikola, P. & Laiho, O. (1962). Mycorrhizal relations in the raw humus layer of northern spruce forests. *Communicationes Instituti Forestalis Fennicae*, **55**, 1–13.

Mikola, P. (1965). Studies on the ectendotrophic mycorrhiza of pine. *Acta Forestalia Fennica*, **79**, 1–56.

Miller, R. M. (1979). Some occurrences of vesicular-arbuscular mycorrhizae in natural and disturbed ecosystems of the Red Desert. *Canadian Journal of Botany*, **57**, 619–23.

Miller, R. M. (1987). The ecology of vesicular-arbuscular mycorrhizae in grass- and shrublands. In *Ecophysiology of VA Mycorrhizal Plants*, ed. G. R. Safir, pp. 135–170. CRC Press, Boca Raton FL.

Miller, R. M., Jarstfer, A. G. & Pilai, J. K. (1987). Biomass allocation in an *Agropyron smithii-Glomus* symbiosis. *American Journal of Botany*, **74**, 114–22.

Miller, R. M., Moorman, T. B. & Schmidt, S. K. (1983). Interspecific plant association effects on vesicular-arbuscular mycorrhiza occurrence in *Atriplex confertifolia*. *New Phytologist*, **95**, 241–6.

Mishustin, E. N. (1967). Mycotrophy in trees and its value in silviculture. In *Mycotrophy in Plants*, ed. A. A. Imshenetskii, pp. 17–34. S. Monson, Jerusalem.

Modjo, H. S. & Hendrix, J. W. (1986). The mycorrhizal fungus *Glomus macrocarpum* as a cause of tobacco stunt disease. *Phytopathology*, **76**, 688–91.

Molina, R. (1981). Ectomycorrhizal specificity in the genus *Alnus*. *Canadian Journal of Botany*, **59**, 325–34.

Molina, R. (ed.) (1985). *Proceedings of the 6th North American Conference on Mycorrhizae*. Forestry Sciences Laboratory, Corvallis, OR.

Molina, R. J. & Trappe, J. M. (1982a). Patterns of ectomycorrhizal host specificity and potential amongst Pacific Northwest conifers and fungi. *Forest Science*, **28**, 423–57.

Molina, R. & Trappe, J. M. (1982b). Lack of mycorrhizal specificity by the ericaceous hosts *Arbutus menziesii* and *Arctostaphylos uva-ursi*. *New Phytologist*, **90**, 495–509.

Moore, J. C., St John, T. V. & Coleman, D. C. (1986). Ingestion of vesicular-arbuscular mycorrhizal hyphae and spores by soil microarthropods. *Ecology*, **66**, 1979–81.

Mosse, B. (1975). Specificity in VA mycorrhizas. In *Endomycorrhizas*, ed. F. E. Sanders, B. Mosse and P. B. Tinker, pp. 469–84. Academic Press, New York.

Mosse, B. & Phillips, J. M. (1971). The influence of phosphate and other nutrients on the development of vesicular-arbuscular mycorrhiza in culture. *Journal of General Microbiology*, **69**, 157–66.

Mosse, B., Stribley, D. P. & LeTacon, F. (1981). Ecology of mycorrhizae and mycorrhizal fungi. *Advances in Microbial Ecology*, **5**, 137–210.

Mueller-Dombois, D. & Ellenberg, H. (1974). *Aims & Methods of Vegetation Ecology*. John Wiley & Sons, New York.

Murakami, Yasuaki (1987). Spatial distribution of *Russula* species in *Castanopsis cuspidata* forest. *Transactions of the British Mycological Society*, **89**, 187–93.

Nadkarni, N. M. (1985). Roots that go out on a limb. *Natural History*, **94**(2), 42–8.

Neil, J. L. Jr (1973). Influence of selected grasses and forbes on soil phosphatase activity. *Canadian Journal of Soil Science*, **53**, 119–21.

Newman, E. I. (1988). Mycorrhizal links between plants: their functioning and ecological significance. *Advances in Ecological Research*, **18**, 243–70.

Newman, E. I., Heap, A. J. & Lawley, R. A. (1981). Abundance of mycorrhizas and root-surface micro-organisms of *Plantago lanceolata* in relation to soil and vegetation: A multi-variate approach. *New Phytologist*, **89**, 95–108.

Newman, E. I. & Reddell, P. (1987). The distribution of mycorrhizas among families of vascular plants. *New Phytologist*, **106**, 745–51.

Nicolson, T. H. (1960). Mycorrhizae in the Gramineae. II. Development in different habitats particularly sand dunes. *Transactions of the British Mycological Society*, **43**, 132–45.

Nicolson, T. H. (1975). Evolution of vesicular-arbuscular mycorrhizas. In *Endomycorrhizas*, ed. F. E. Sanders, B. Mosse and P. B. Tinker, pp. 25–34. Academic Press, New York.

Nicolson, T. H. & Johnston, C. (1979). Mycorrhiza in the Graminae. II. *Glomus fasciculatus* as the endophyte of pioneer grasses in a maritime sand dune. *Transactions of the British Mycological Society*, **72**, 261–8.

Niederpruem, D. J. (1965). Carbohydrate metabolism. 2. Tricarboxylic acid cycle. In *The Fungi, Volume1, The Fungal Cell*, ed. G. C. Ainsworth and A. S. Sussman, pp. 269–300. Academic Press New York.

Norkrans, B. (1950). Studies in growth and cellulolytic enzymes of Tricholoma. *Symbolae Botanicae Uppsaliensis*, **11**, 1–126.

Nye, P. H. & Tinker, P. B. (1977). *Solute Movement in the Soil–Root System*. University of California Press.

Ocampo, J. A. & Hayman, D. A. (1981). Influence of plant interactions on vesicular-arbuscular mycorrhizal infections. II. Crop rotations and residual effects of non-host plants. *New Phytologist*, **87**, 333–43.

Ocampo, J. A., Martin, J. & Hayman, D. S. (1980). Influence of plant interactions on vesicular-arbuscular mycorrhizal infections. I. Host and non-host plants grown together. *New Phytologist*, **84**, 27–35.

Odum, E. P. (1969). The strategy of ecosystem development. *Science*, **164**, 262–70.

Ogawa, M. (1985). Ecological characters of ectomycorrhizal fungi and their mycorrhizae: an introduction to the ecology of higher fungi. *JARQ*, **18**, 305–14.

Ogawa, M., Fujita, H., Ishizuka, K. & Yambe, Y. (1983). Controlling of mycorrhizal fungi in pine forest. *Third International Mycological Congress, Tokyo*, p. 218 (abstract).

Ogawa, M., Yambe, Y. & Ishizuka, K. (1981). Fungal and soil microbial flora in the natural stand of *Fagus crenata* and *Fagus japonica*. *Bulletin of the Forest Production Research Institute*, **314**, 71–88.

Oliver, A. J., Smith, S. E., Nicholas, D. J. D., Wallace, W. & Smith, P. S. (1983). Activity of nitrate reductase in *Trifolium subterraneum*: Effects of mycorrhizal infection and phosphate nutrition. *New Phytologist*, **94**, 63–79.

O'Neill, E. G. & Norby, R. J. (1988). Differential responses of ecto- and endomycorrhizae to elevated atmospheric CO_2. *Bulletin of the Ecological Society of America (supplement)*, **69**, 248–9.

O'Neill, E. G., Rogers, H. H. & Prior, S. A. (1989). Rhizosphere microbiology and plant growth responses in field-grown cotton exposed to elevated CO_2 via a free air CO_2 enrichment (FACE) array. *Bulletin of the Ecological Society of America (supplement)*, **70**, 219.

Owusu-Bennoah, E. & Wild, A. (1979). Autoradiography of the depletion zone of phosphate around onion roots in the presence of vesicular arbuscular mycorrhiza. *New Phytologist*, **82**, 133–40.

Owusu-Bennoah, E. & Wild, A. (1980). Effects of vesicular-arbuscular mycorrhiza on the size of the labile pool of soil phosphate. In *Tropical Mycorrhiza Research*, ed. P. Mikola, p. 231. Clarendon Press, Oxford.

Pacovsky, R. S., Fuller, G. & Paul, E. A. (1985). Influence of soil on the interactions between endomycorrhizae and *Azospirillum* in sorghum. *Soil Biology and Biochemistry*, **17**, 525–31.

Pang, P. C. & Paul, E. A. (1980). Effects of vesicular-arbuscular mycorrhiza on carbon-^{14} & nitrogen-^{15} distribution in nodulated fababeans. *Canadian Journal of Soil Science*, **60**, 241–50.

Parrish, J. A. D. & Bazzaz, F. A. (1976). Underground niche separation in successional plants. *Ecology*, **57**, 1281–8.

Paul, E. A. & Kucey, R. M. N. (1981). Carbon flow in plant microbial associations. *Science*, **213**, 473–4.

Paulitz, T. C. & Menge, J. A. (1986). The effects of a mycoparasite on the mycorrhizal fungus, *Glomus deserticola*. *Phytopathology*, **76**, 351–4.

Pearson, V. & Read, D. J. (1973). The biology of mycorrhiza in the Ericaceae. II. The transport of carbon and phosphorus by the endophyte and the mycorrhiza. *New Phytologist*, **72**, 1325–31.

Pendelton, R. L. & Smith, B. N. (1983). Vesicular-arbuscular mycorrhizae of weedy and colonizer plant species at disturbed sites in Utah. *Oecologia* (*Berlin*), **59**, 296–301.

Perry, D. A., Amaranthus, M. P., Borchers, J. G., Borchers, S. L. & Brainerd, R. E. (1989). Bootstrapping in ecosystems. *BioScience*, **39**, 230–7.

Pickett, S. T. A. & White, P. S. (ed.) (1985). *The Ecology of Natural Disturbance and Patch Dynamics*. Academic Press, Orlando, FL.

Pirozynski, K. A. (1976). Fossil fungi. *Annual Review of Phytopathology*, **14**, 237–46.

Pirozynski, K. A. (1981). Interactions between fungi and plants through the ages. *Canadian Journal of Botany*, **59**, 1824–7.

Pirozynski, K. A. & Malloch, D. W. (1975). The origins of land plants: a matter of mycotrophism. *Biosystems*, **6**, 153–64.

Pond, E. C., Menge, J. A. & Jarrell, W. M. (1984). Improved growth of tomato in salinized soil by vesicular-arbuscular mycorrhizal fungi collected from saline soils. *Mycologia*, **76**, 74–84.

Ponder, F. Jr (1980). Rabbits and grasshoppers: vectors of endomycorrhizal fungi on new coal mine spoil. *North Central Forest Experiment Station Research Note NC*, **250**, 1–2.

Powell, C. L. (1979). Spread of mycorrhizal fungi through soil. *New Zealand Journal of Agricultural Research*, **22**, 335–9.

Powell, C. L. (1980). Mycorrhizal infectivity of eroded soils. *Soil Biology and Biochemistry*, **12**, 247–57.

Powell, C. L. (1984). Field inoculation with VA mycorrhizal fungi. In *VA Mycorrhiza*, ed. C. L. Powell and D. J. Bagyaraj, pp. 205–22. CRC Press, Boca Raton, FL.

Puppi, G. (1983). Vesicular-arbuscular mycorrhizae and biological nitrogen fixation. A Review. *Micol. Ital.*, **12**, 3–10.

Rabatin, S. C. & Stinner, B. R. (1985). Anthropods as consumers of vesicular-arbuscular mycorrhizal fungi. *Mycologia*, **77**, 320–2.

Ratnayake, M., Leonard, R. T. & Menge, J. A. (1978). Root exudation in relation to supply of phosphorus and its possible relevance to mycorrhizal formation. *New Phytologist*, **81**, 543–52.

Rayner, A. D. M., Coates, D., Ainsworth, A. M., Adams, T. J. H., Williams, E. N. D. & Todd, N. K. (1984). The biological consequences of the individualistic mycelium. In *The Ecology and Physiology of the Fungal Mycelium*, ed. D. H. Jennings and A. D. M. Rayner, pp. 509–40. Cambridge University Press, Cambridge.

Read, D. J. (1983). the biology of mycorrhiza in the Ericales. *Canadian Journal of Botany*, **61**, 985–1004.

Read, D. J. (1984). The structure and function of the vegetative mycelium of mycorrhizal roots. In *The Ecology and Physiology of the Fungal Mycelium*, ed. D. H. Jennings and A. D. M. Rayner, pp. 215–40. Cambridge University Press, Cambridge.

Read, D. J. & Bajwa, R. (1985). Some nutritional aspects of the biology of ericacious mycorrhizas. *Proceedings of the Royal Society of Edinburgh*, **85B**, 317–32.

Read, D. J., Francis, R. & Finlay, R. D. (1985). Mycorrhizal mycelia and nutrient cycling in plant communities. In *Ecological Interactions in Soil*, ed. A. H. Fitter, pp. 193–217. Blackwell Scientific Publications, Oxford.

Read, D. J. & Haselwandter, K. (1981). Observations on the mycorrhizal status of some alpine plant communities. *New Phytologist*, **88**, 341–52.

Read, D. J., Koucheki, H. K. & Hodgson, J. (1976). Vesicular-arbuscular mycorrhiza in natural vegetation systems. *New Phytologist*, **76**, 641–53.

Reece, P. E. & Bonham, C. D. (1978). Frequency of endomycorrhizal infection in grazed and ungrazed bluegrama plants. *Journal of Range Management*, **31**, 149–51.

Reeves, F. B., Wagner, D. W., Moorman, T. & Kiel, J. (1979). The role of endomycorrhizae in revegetation practices in the semi-arid west. I. A comparison of incidence of mycorrhizae in severely disturbed vs. natural environments. *American Journal of Botany*, **66**, 1–13.

Reid, C. P. P., Kidd, F. A. & Ekwebelam, S. A. (1983). Nitrogen nutrition, photosynthesis and carbon allocation in ectomycorrhizal pine. *Plant and Soil*, **71**, 415–32.

Reid, C. P. P. & Woods, F. W. (1969). Translocation of C^{14}-labeled compounds in mycorrhizae and its implications in interplant nutrient cycling. *Ecology*, **50**, 179–81.

Reis, E. M., Cook, R. J. & McNeal, B. L. (1982). Effect of mineral nutrition on take-all of wheat. *Phytopathology*, **72**, 224–9.

Rhodes, L. H. & Gerdemann, J. W. (1975). Phosphate uptake zones of mycorrhizal and non-mycorrhizal onions. *New Phytologist*, **75**, 555–61.

Risser, P. G., Birney, E. C., Blocker, H. D., May, S. W., Parton, W. J. & Weins, J. A. (1981). *The True Prairie Ecosystem*. Hutchinson Ross Publishing Company, Stroudsburg, PA.

Ritz, K. & Newman, E. I. (1984). Movement of phosphorus-32 between intact grassland plants of the same age. *Oikos*, **43**, 138–142.

Ritz, K. & Newman, E. I. (1985). Evidence for rapid cycling of phosphorus from dying roots to living plants. *Oikos*, **45**, 174–80.

Rommell, L. G. (1938). A trenching experiment in spruce forest and its bearing on the problems of mycotrophy. *Svensk Botanisk Tidskrift*, **32**, 89–99.

Rommell, L. G. (1939). The ecological problem of mycotrophy. *Ecology*, **20**, 163–7.

Rose, S. L., Perry, D. A., Pilz, D. & Schoenenberger, M. M. (1983). Allelopathic effects of litter on the growth and colonization of mycorrhizal fungi. *Journal of Chemical Ecology*, **9**, 1153–62.

Rose, S. L. & Youngberg, C. T. (1981). Tripartite associations in snowbrush (*Leanothus velutinus*): effect of vesicular-arbuscular mycorrhizae on growth, nodulation and nitrogen fixation. *Canadian Journal of Botany*, **59**, 34–9.

Ross, J. P. & Ruttencutter, R. (1977). Population dynamics of two vesicular-arbuscular endomycorrhizal fungi and the role of hyperparasitic fungi. *Phytopathology*, **67**, 490–6.

Rothwell, F. M. & Holt, C. (1978). Vesicular-arbuscular mycorrhizae established with *Glomus fasciculatus* spores isolated from the feces of cricetine mice. *USDA Forest Service Research Note* NE-259.

Roughgarden, J. (1979). *Theory of Population Genetics and Evolutionary Ecology: an Introduction*. MacMillan Publ. Co., New York.

Safir, G. R. (ed.) (1987). *Ecophysiology of VA Mycorrhizal Plants*. CRC Press, Boca Raton, FL.
Saif, S. R. & Khan, A. G. (1977). The effect of vesicular-arbuscular mycorrhizal associations on growth of cereals. III. Effects of barley growth. *Plant and Soil*, **47**, 17–26.
Salick, J., Herra, R. & Jordan, C. F. (1983). Termitaria: Nutrient patchiness in nutrient-deficient rain forests. *Biotropica*, **15**, 1–7.
Sanders, F. E. & Tinker, P. B. (1971). Mechanism of absorption of phosphate from soil by *Endogone* mycorrhizas. *Nature* (*London*), **233**, 278–9.
Sanders, F. E. & Tinker, P. B. (1973). Phosphate flow into mycorrhizal roots. *Pesticide Science*, **4**, 385–95.
Schemakhanova, N. M. (1962). *Mycotrophy of Woody Plants*. S. Monson, Binding: Wiener Bindery, Ltd, Jerusalem.
Schenck, N. C. (1981). Can mycorrhizae control root disease? *Plant Disease*, **65**, 230–4.
Schenck, N. C. (1982). *Methods and Principles of Mycorrhizal Research*. American Phytopathological Society, St Paul, MN.
Schenck, N. C. & Perez, Y. (1988). Manual for the identification of VA mycorrhizal fungi. INVAM, Gainesville, FL.
Schimpf, D. J., Henderson, J. A. & MacMahon, J. A. (1980). Some aspects of succession in the spruce–fir forest zone of northern Utah. *Great Basin Naturalist*, **40**, 1–26.
Schmidt, S. K. & Reeves, F. B. (1984). Effect of non-mycorrhizal pioñeer plant *Salsola kali* (Chenopodiaceae) on vesicular-arbuscular mycorrhizal fungi. *American Journal of Botany*, **71**, 1035–9.
Schmidt, S. K. & Scow, K. M. (1986). Mycorrhizal fungi on the Galapagos islands. *Biotropica*, **18**, 236–40.
Schramm, J. R. (1966). Plant colonizing studies on black wastes from anthracite mining in Pennsylvania. *Transactions of the American Philosophical Society*, **47**, 1–331.
Schoknecht, J. O. & Hattingh, M. J. (1976). X-ray microanalysis of elements of VA-mycorrhizal and non-mycorrhizal onions. *Mycologia*, **68**, 296–303.
Schwab, S. & Reeves, F. B. (1981). The role of endomycorrhizae in revegetation practices in the semi-arid west. III. Vertical distribution of vesicular-arbuscular (VA) mycorrhiza inoculum potential. *American Journal of Botany*, **68**, 1293–7.
Shuster, R. A. & Bye, R. A., Jr (1983). Patterns of variation in exotic races of maize (*Zea mays*: Gramineae) in a new geographic area. *Journal of Ethnobiology*, **3**, 157–74.
Simkin, T. & Fiske, R. S. (1983). *Krakatau 1883 – the Volcanic Eruption and its Effects*. Smithsonian Institution Press, Washington, DC.
Singer, R., Araujo, L. de J., & Da, S. (1979). Litter decomposition and ectomycorrhizae in Amazonian forests. I. A comparison of litter decomposing and ectomycorrhizal basidiomycetes in qatosol-terra-firme rain forest and white podsol. Campinarana. *Acta Amazonica*, **9**, 25–41.
Singh, K. & Varma, A. K. (1981). Endogonaceous spores associated with xerophytic plants in northern India. *Transactions of the British Mycology Society*, **77**, 655–8.
Skujins, J. & Allen, M. F. (1986). Use of mycorrhizae for land rehabilitation. *MIRCEN Journal*, **2**, 161–76.
Slankis, V. (1948). Einfluss von Exudaten von *Boletus variegatus* auf die dichotomische verzweigung isolierter kiefern wurzeln. *Physiologia Plantarum*, **1**, 390–400.
Slankis, V. (1951). Uber den einfluss von 8-indolylessigsaure und anderen wuchsstoffen auf das wachstum von kiefernwuryeln. I. *Symbolae Botanicaae Uppsalienses*, **11**, 1–63.
Slankis, V. (1973). Hormonal relationships in mycorrhiza. In *Ectomycorrhizae: their Ecology and Physiology*, ed. G. C. Marks and T. T. Kozlowski, pp. 213–98. Academic Press, New York.

Sollins, P., Grier, C. C., McCorison, F. M., Cromack, K., Jr, Fogel, R. & Fredriksen, R. L. (1980). The internal element cycles of an old-growth Douglas-fir stand in Oregon. *Ecological Monographs*, **50**, 261–85.

Sondergaard, M. & Laegaard, S. (1977). Vesicular-arbuscular mycorrhiza in some aquatic plants. *Nature (London)*, **268**, 232–3.

Sparling, G. P. & Tinker, P. B. (1975). Mycorrhizas in Pennine Grassland. In *Endomycorrhizas*, ed. F. E. Sanders, B. Mosse and P. B. Tinker, pp. 545–60. Academic Press, New York.

Sparling, G. P. & Tinker, P. B. (1978). Mycorrhizal infection in Pennine grassland. I. Levels of infection in the field. *Journal of Applied Ecology*, **15**, 943–50.

Staffeldt, E. E. & Vogt, K. B. (1975). Mycorrhizae of desert Biome Research Memo. 75–37. Utah State University, Logan, UT, 7pp.

Stahl, E. (1900). Der Sinn der mycorrhizenbildung. *Jahrbucher für wissenschaftliche Botanik*, **34**, 539–668.

Stahl, P. O. & Christensen, M. (1983). Mycorrhizal fungi associated with *Bouteloua* and *Agropyron* in Wyoming sagebrush-grass lands. *Mycologia*, **74**, 877–85.

Stahl, P. O. & Smith, W. K. (1984). Effects of different geographic isolates of *Glomus* on the water relations of *Agropyron smithii. Mycologia*, **76**, 261–7.

Stalhandski, C., Svensson, C. & Sarnstrand. (1977). Chloromycorrhizin A. *Acta Crystallographica* **B33**, 870–3.

Stanton, N. L., Allen, M. F., & Campion, M. (1981). The effect of the pesticide carbofuran on soil organisms and root and shoot production in shortgrass prairie. *Journal of Applied Ecology*, **18**, 417–31.

Stark, N. M. (1972). Nutrient cycling pathways and litter fungi. *BioScience*, **22**, 355–60.

Stark, N. M. & Jordan C. F. (1978). Nutrient retention by the root mat of an Amazonian rain forest. *Ecology*, **59**, 434–7.

States, J. (1984). Hypogeous mycorrhizal fungi associated with ponderosa pine: sporocarp phenology. In *Proceedings of the 6th North American Conference on Mycorrhizae*, ed. R. Molina, p. 271. Forest Research Laboratory, Corvallis, OR (abstract).

Stebbins, G. L. (1974). *Flowering Plants, Evolution Above the Species Level.* Harvard University Press, Cambridge, MA.

St. John, T. V. (1980a). A survey of mycorrhizal infection in an Amazonian rain forest. *Acta Amazonica*, **10**, 527–33.

St John, T. V. (1980b). Root size, root hairs, and mycorrhizal infection: a re-examination of Baylis's hypothesis with tropical trees. *New Phytologist*, **84**, 483–7.

St John, T. V. & Coleman, D. C. (1983). The role of mycorrhizae in plant ecology. *Canadian Journal of Botany*, **61**, 1005–14.

St John, T. V., Coleman, D. C. & Reid, C. P. P. (1983a). Association of vesicular-arbuscular mycorrhizal hyphae with soil organic particles. *Ecology*, **64**, 957–959.

St John, T. V., Coleman, D. C. & Reid, C. P. P. (1983b). Growth and spatial distribution of nutrient absorbing organs: selective exploitation of soil heterogeneity. *Plant and Soil*, **71**, 487–93.

St John, T. V. & Hunt, H. W. (1983). Statistical treatment of VAM infection data. *Plant and Soil*, **73**, 307–13.

St John, T. V. & Koske, R. E. (1988). Statistical treatment of Endogonaceous spore counts. *Transactions of the British Mycological Society*, **91**, 117–21.

Stribley, D. P. & Read, D. J. (1974). The biology of mycorrhizae in the Ericaceae. IV. The Effect of mycorrhizal infection on the uptake of ^{15}N from labelled soil by *Vaccinium macrocarpon* Ait. *New Phytologist*, **73**, 1449–55.

Stubblefield, S. P. & Banks, H. P. (1983). Fungal remains in the Devonian trimerophyte *Psilophyton dawsonii. American Journal of Botany*, **70**, 1258–61.

Stubblefield, S. P. & Taylor, T. N. (1988). Tansley Review No. 12. Recent advances in palaeomycology. *New Phytologist*, **108**, 3–25.

Stubblefield, S. P., Taylor, T. N. & Trappe, J. M. (1987a). Fossil mycorrhizae: a case for symbiosis. *Science*, **237**, 59–60.

Stubblefield, S. P., Taylor, T. N. & Trappe, J. M. (1987b). Vesicular-arbuscular mycorrhizae from the Triassic of Antarctica. *American Journal of Botany*, **47**, 1904–11.

Sutton, J. C. & Sheppard, B. R. (1976). Aggregation of sand-dune soil by endomycorrhizal fungi. *Canadian Journal of Botany*, **54**, 326–33.

Sylvia, D. M. (1986). Spatial and temporal distribution of vesicular-arbuscular mycorrhizal fungi associated with *Uniola paniculata* in Florida foredunes. *Mycologia*, **78**, 728–34.

Sylvia, D. M., Hung, L. L. & Graham, J. H. (1987). *Mycorrhizae in the Next Decade, Practical Applications and Research Priorities*. IFAS, Gainesville, FL.

Szaniszlo, P. J., Powell, P. E., Reid, C. P. P. & Cline, G. R. (1981). Production of hydroxamate siderophore iron chelators by ectomycorrhizal fungi. *Mycologia*, **73**, 1158–74.

Tarasova, Z. H. G. & Dumikyan, A. D. (1984). The formation of mycorrhiza and the productivity altitude. *Biologicheskii Zhurnal Armenii*, **37**, 275–81.

Thaxter, R. (1922). A revision of the Endogonaceae. *Proceedings of the American Academy of Arts and Sciences*, **57**, 291–351.

Tobiessen, P. & Werner, M. B. (1980). Hardwood seedling survival under plantations of scotch pine and red pine in central New York. *Ecology*, **61**, 25–9.

Tommerup, I. C. (1984). Development of infection by a vesicular-arbuscular mycorrhizal fungus in *Brassica napus* L. and *Trifolium subterraneum* L. *New Phytologist*, **98**, 487–95.

Trappe, J. M. (1977). Selection of fungi for ectomycorrhizal inoculation in nurseries. *Annual Review of Phytopathology*, **15**, 203–22.

Trappe, J. M. (1980). Truffles in North America. *McIlvainae*, 1–5.

Trappe, J. M. (1981). Mycorrhizae and productivity of arid and semi-arid rangelands. In *Advances in Food Producing Systems for Arid and Semi-arid Lands*, ed. J. T. Manassah and E. J. Briskey, pp. 581–99. Academic Press, New York.

Trappe, J. M. (1987). Phylogenetic and ecological aspects of mycotrophy in the Angiosperms from an evolutionary standpoint. In *Ecophysiology of VA Mycorrhizal Plants*, ed. G. R. Safir, pp. 5–25. CRC Press, Boca Raton, FL.

Trappe, J. M. (1988). Lessons from alpine fungi. *Mycologia*, **80**, 1–10.

Trappe, J. M. & Fogel, R. D. (1977). Ecosystematic functions of mycorrhizae. In *The Below-ground Ecosystem*, ed. J. K. Marshall, pp. 205–14. Range Science Department Science Series, Colorado State University, Fort Collins, CO.

Trappe, J. M. & Maser, C. (1976). Germination of spores of *Glomus macrocarpus* (Endogonaceae) after passage through a rodent digestive tract. *Mycologia*, **68**, 433–6.

Trent, J. D., Svejcar, T. J. & Christensen, S. (1989). Effects of fumigation on growth, photosynthesis, water relations and mycorrhizal development of winter wheat in the field. *Canadian Journal of Plant Science*, **69**, 535–40.

Trojanowski, J., Haider, K. & Huettermann, A. (1984). Decomposition of carbon-14-labeled lignin, holocellulose and lignocellulose by mycorrhizal fungi. *Archives of Microbiology*, **139**, 202–6.

Vandermeer, J., Hazlett, B. & Rathcke, B. (1985). Indirect facilitation and mutualism. In *The Biology of Mutualism, Ecology and Evolution*, ed. E. H. Boucher, pp. 326–43. Oxford University Press, New York.

Van Kessel, C., Singleton, P. W. & Hoben, H. J. (1985). Enhanced nitrogen-transfer from a soybean to maize by vesicular-arbuscular mycorrhizal fungi. *Plant Physiology*, **79**, 562–3.

Virginia, R. A., Jenkins, M. B. & Jarrell, W. M. (1986). Depth of root symbiont occurrence in soil. *Biology and Fertility of Soils*, **2**, 127–30.

Vogt, K. A. & Edmonds, R. L. (1980). Patterns of nutrient concentration in basidiocarps in western Washington. *Canadian Journal of Botany*, **58**, 694–8.

Vogt, K. A., Edmonds, R. L. & Grier, C. C. (1981). Biomass and nutrient concentrations of sporocarps produced by mycorrhizal and decomposer fungi in *Abies amabilis* stands. *Oecologia* (*Berlin*), **56**, 170–5.

Vogt, K. A., Grier, C. C., Meier, C. E. & Edmonds, R. L. (1982). Mycorrhizal role in net production and nutrient cycling in *Abies amabilis* ecosystems in western Washington. *Ecology*, **63**, 370–80.

Vogt, K. A., Grier, C. C. & Vogt, D. J. (1986). Production, turnover, and nutrient dynamics of above and below-ground detritus of world forests. *Advances in Ecological Research*, **15**, 303–77.

Waaland, M. E. & Allen, E. B. (1987). Relationships between VA mycorrhizal fungi and plant cover following surface mining in Wyoming. *Journal of Range Management*, **40**, 271–6.

Wagner, C. A. & Taylor, T. N. (1981). Evidence for endomycorrhizae in Pennsylvanian age plant fossils. *Science*, **212**, 562–3.

Walker, C. (1985). Taxonomy of the Endogonaceae. In *Proceedings of the 6th North American Conference on Mycorrhizae*, ed. R. Molina, pp. 193–8. Forest Research Laboratory, Corvallis, OR.

Wallace, L. L. (1981). Growth, morphology and gas exchange of mycorrhizal and nonmycorrhizal *Panicum coloratum* L., a C_4 grass species, under different clipping and fertilization regimes. *Oecologia* (*Berlin*), **49**, 272–8.

Wallace, L. L. (1987a). Mycorrhizas in grasslands: interactions of ungulates, fungi and drought. *New Phytologist*, **105**, 619–32.

Wallace, L. L. (1987b). Effects of clipping and soil compaction on growth, morphology and mycorrhizal colonization of *Schizachyrium scoparium*, a C_4 bunchgrass. *Oecologia* (*Berlin*), **72**, 423–8.

Warcup, J. H. (1981). The mycorrhizal relationships of Australian orchids. *New Phytologist*, **87**, 371–81.

Warner, A. & Mosse, B. (1980). Independent spread of vesicular-arbuscular mycorrhizal fungi in soil. *Transactions of the British Mycological Society*, **74**, 407–10.

Warner, N. J. (1985). Dispersal of vesicular-arbuscular mycorrhizal fungi in a disturbed arid ecosystem: the potential roles of biotic and abiotic agents. MS thesis, Utah State University, Logan, UT.

Warner, N. J., Allen, M. F. & MacMahon, J. A. (1987). Dispersal agents of vesicular-arbuscular mycorrhizal fungi in a disturbed arid ecosystem. *Mycologia*, **79**, 721–30.

Warnock, A. J., Fitter, A. H. & Usher, M. B. (1982). The influence of a springtail *Folsomia candida* (Insecta: Collembola) on the mycorrhizal association of leek *Allium porrum* and the vesicular-arbuscular mycorrhizal endophyte *Glomus fasciculatum*. *New Phytologist*, **90**, 285–92.

Watling, R. (1981). Relationships between macromycetes and the development of higher plant communities. In: *The Fungal Community*, ed. D. T. Wicklow and G. C. Carroll, pp. 427–58. Marcel Dekker, Inc., New York.

Watling, R. (1988). Larger fungi and some of the earth's catastrophies. *Proceedings of the Royal Society of Edinburgh*, **94B**, 49–60.

Weaver, J. E. & Albertson, F. W. (1943). Resurvey of grasses, forbs, and underground plant parts at the end of the Great Drought. *Ecological Monographs*, **13**, 63–117.

Weijman, A. C. M. & Meuzelaar, L. C. (1979). Biochemical contributions to the taxonomic status of the Endogonaceae. *Canadian Journal of Botany*, **572**, 284–91.

Weins, J. A. (1977). On competition and variable environments. *American Scientist*, **65**, 590–7.

Weins, J. A., Crawford, C. S. & Gosz, J. R. (1986). Boundary dynamics: a conceptual framework for studying landscape ecosystems. *Oikos*, **45**, 421–7.

Weiss, F. E. (1904). A mycorrhiza from the lower coal measures. *Annals of Botany*, **18**, 255–67.

Went, F. W. & Stark, N. (1968). The biological and mechanical role of soil fungi. *Proceedings of the National Academy of Sciences, USA*, **60**, 4479–504.

West, D. C., Shugart, H. H. & Botkin, D. B. (1981). *Forest Succession. Concepts and Application*. Springer-Verlag, New York.

West, N. E. & Van Pelt, N. S. (1987). Successional patterns in pinyon–juniper woodlands. *USDA Forest Service Intermountain Research Station General Technical Report INT-215*, pp. 43–52.

White, D. P. (1941). Prairie soil as a medium for tree growth. *Ecology*, **22**, 398–407.

White, J. (1984). Plant metamerism. In: *Perspectives on Plant Population Ecology*, ed. R. Dirzo and J. Sarukhan, pp. 15–47. Sinauer Associates Inc., Sunderland, MA.

Whittaker, R. H., Levin, S. A. & Root, R. B. (1973). Niche, habitat, and ecotope. *American Naturalist*, **107**, 321–38.

Whittingham, J. & Read, D. J. (1982). Vesicular-arbuscular mycorrhiza in natural vegetation systems III. Nutrient transfer between plants with mycorrhizal interconnections. *New Phytologist*, **90**, 277–84.

Wilcox, H. E. (1982). Morphology and development of ecto- and ectendomycorrhizae. In *Methods and Principles of Mycorrhizal Research*, ed. N. C. Schenck, pp. 103–13. American Phytopathological Society, St Paul, MN.

Wilde, S. A., Corey, R. B. & Iyer, J. G. (1979). Mycorrhiza-related terminology. *Forestry Research Notes*, **223**, 1–5.

Williams, G. C. (1966). *Adaptation and Natural Selection*. Princeton University Press, Princeton, NJ.

Williams, S. E. & Aldon, E. F. (1976). Endomycorrhizal (vesicular-arbuscular) associations of some arid zone shrubs. *The Southwestern Naturalist*, **20**, 437–44.

Williamson, M. (1972). *The Analysis of Biological Populations*. Academic Press, London.

Wilson, D. S. (1983). The effect of population structure on the evolution of mutualism: a field test involving burying beetles and their phoretic mites. *American Naturalist*, **121**, 851–70.

Wilson, J. M. (1984). Competition for infection between vesicular-arbuscular mycorrhizal fungi. *New Phytologist*, **97**, 427–35.

Wilson, J. M., Trinick, M. J. & Parker, C. A. (1983). The identification of vesicular-arbuscular mycorrhizal fungi using immunofluorescence. *Soil Biology and Biochemistry*, **15**, 439–45.

Wood, T. (1984). Commercialization of VA mycorrhizal inoculum: The reclamation market. In *Proceedings of the Conference on VA Mycorrhizae and Reclamation of Arid and Semi-Arid Lands*, ed. S. E. Williams and M. F. Allen, pp. 21–27. University of Wyoming Agricultural Experiment Station Scientific Report No. SA1261, Laramie, WY.

Wood, T., Bormann, F. H. & Voight, G. K. (1984). Phosphorus cycling in a northern hardwood forest, biological and chemical control. *Science*, **223**, 391–3.

Woods, F. W. & Brock, K. (1964). Interspecific transfer of Ca^{45} and P^{32} by root systems. *Ecology*, **45**, 886–9.

Woolhouse, H. W. (1975). Membrane structure and transport problems considered in relation to phosphorus and carbohydrate movements and the vegetation of endotrophic mycorrhizal associations. In *Endomycorrhizas*, ed. F. E. Sanders, B. Mosse and P. B. Tinker, pp. 209–40. Academic Press, London.

Wright, S. F., Morton, J. B. & Sworobuk, J. E. (1987). Identification of a vesicular-arbuscular mycorrhizal fungus by using monoclonal antibodies in an enzyme-linked immunosorbent assay. *Applied and Environmental Microbiology*, **53**, 2222–5.

Yocum, D. H. (1983). The costs and benefits to plants forming mycorrhizal associations. PhD dissertation, State University of New York at Stony Brook.

Yocum, D. H. (1985). VA mycorrhizae and host plant reproduction: a study with green peppers. In *Proceedings of the 6th North American Conference on Mycorrhizae*, ed. R. Molina. Forest Research Laboratory. Corvallis, OR (abstract).

Yocum, D. H., Boosalis, M. G., Flessner, T. R. & Cunningham, E. A. (1987). The effects of prolonged drought on the yield of mycorrhiza and nonmycorrhizal grain crops. In *Mycorrhizae in the Next Decade*, ed. D. M. Sylvia, L. L. Hung and J. H. Graham. IFAS, Gainesville, FL (abstract).

Zajicek, J. M., Hetrick, B. A. Daniels & Owensby, C. E. (1986). The influence of soil depth on mycorrhizal colonization of forbs in the tall-grass prairie. *Mycologia*, **78**, 316–20.

Zambonelli, A. & Govi, G. (1983). Micorrizazione in semenzaio di *Quercus pubescens* Willd. con specie del genere *Tuber*. *Mic. Ital.* **1**m 17–22.

Zambonelli, A. & Morara, M. (1984). Le specie di *Tuberales* dell'Emilia-Romagna: ecologia e distribuzione. *Estratto da natura e montagna*, **4**, 9–32.

Index